GENTLEMEN
PREFERRED
DRY FLIES

GENTLEMEN PREFERRED DRY FLIES

THE DRY FLY AND THE NYMPH, EVOLUTION AND CONFLICT

WILLIAM C. BLACK

UNIVERSITY OF NEW MEXICO PRESS

ALBUQUERQUE

© 2010 by the University of New Mexico Press
All rights reserved. Published 2010
Printed in the United States of America
16   15   14   13   12   11   10   1   2   3   4   5   6   7

Library of Congress Cataloging-in-Publication Data

Black, William C., 1931–
    Gentlemen preferred dry flies : the dry fly and the nymph,
    evolution and conflict / William C. Black III.
        p. cm.
Includes bibliographical references.

ISBN 978-0-8263-4795-4 (pbk. : alk. paper)
1. Fishers—Biography.
2. Fly tiers—Biography.
3. Fly fishing—History.
I. Title.

SH414.B58 2010
799.12'409—dc22
                                        2010000874

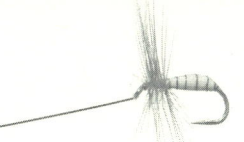

TABLE OF CONTENTS

INTRODUCTION
CASTE AMONGST FLY FISHERS?

*M*r. Gordon was greatly admired by the regulars at our golf club. Well into his seventies, he could pretty consistently "shoot his age," often lower. That's no mean feat on a 7,000-plus yard course with narrow fairways lined by lush, green rough and shaded by long-limbed elms. I was playing for my college team at the time, and there were days "in the trees" when Mr. Gordon would have closed me out early. He was also a renowned fly fisherman and a member of the historic Wigwam Club on the South Platte River west of Denver. When I once asked him about his fishing, I was solemnly advised that a true sportsman fishes the dry fly *only* and a gentleman casts *only* to rising trout. He explained that deceiving a wary trout by successfully imitating a delicate insect made the pursuit a fair one, especially when the deception occurred at the very instant when the trout's feeding instincts were intently focused. Striding the fairways, tall and straight, attired in traditional golfing knickers and stockings, Mr. Gordon's fishing ethic sounded much like his golf game, precise, controlled and efficient, no balls straying off the fairway, no "three putts". I accepted his ethic as a quaint old-school-tie sort of thing, analogous to hunting to the hounds. It was the early 1950s. I'd been whoring around with the tawdry new spinning lures, and I felt a bit ashamed and puzzled too. My father and my Uncle Charlie, together with a half-dozen of their friends, fished a brace of wet flies downstream, swing across, two-steps down, and repeat. Not

I

very sporting, I thought. Yet they were well-respected citizens, paying taxes, voting, serving jury duty and I couldn't recall any arrests. Heck, Charlie was a graduate of MIT! Was there really a caste system among fly fishermen, the Brotherhood of the Angle, where good fellowship had been the byword since Sir Izaak's day? I'd thought that fishers of the fly were hopelessly congenial, intent upon slapping backs, swapping lies, and forming clubs. Could the infinitesimal surface film, the meniscus that divides the dry world from the wet, divide us? Admittedly one side faced heavenward, the other, the lower world where wet flies were fished, and even bait. After graduation, as more practical matters intervened, I put the matter aside.

Boorishly, now in *my* seventies, I often fish nymphs and streamers and have gotten over feeling bad about myself, at least in context. Yet I had great respect for Mr. Gordon and never overcame a gnawing curiosity about when and how dry fly elitism developed. Accordingly, when a friend gave me a copy of Glen Law's excellent *A Concise History of Fly Fishing* (2003), I felt compelled to finally explore "purism." I learned that the issue played out around the beginning of the twentieth century in the form of a battle of words between a pair of English protagonists. One gentleman, F. M. Halford, essentially defined and developed Mr. Gordon's ethic. Meanwhile, G. E. M. Skues, Halford's contemporary, and also a gentleman, placed the "wet fly" nymph on a dais of equal eminence. But it wasn't a private skirmish. Sides were taken, the popular vote going so heavily to Halford that the nymph was disallowed on certain waters! Now I was *really* curious. After all, Mr. Gordon was still flying the Halford banner on the other side of the Atlantic some fifty years later. Recent biographies were easy to find. Tony Hayter's *F. M. Halford and the Dry-Fly Revolution* (2002) dovetailed with Kenneth Robson's *The Essential G. E. M. Skues* (1998). It wasn't enough, though; now I needed to better understand the evolution of the dry fly and the nymph and, more to the point, how the consequences of the debate translated across time to twentieth-century America. The idea of trying to identify our early-day anglers, some of whom had corresponded with or actually knew Halford and Skues, was appealing, since these men would logically have been important in developing our own styles and preferences. I turned to works by a half-dozen professional historians in an effort to identify those anglers whose signal contributions brought us from the past to the present. There was a decent collection of American titles dating back into the 1940s on my shelves, so rather than going the library route, I decided to try to acquire the earlier works. Why? It's the difference between reading another's interpretation of a book and listening to the author's words firsthand. Besides, I can never extract the information I need the first time through. Thanks to the Internet, it has been relatively easy to acquire these titles. As long as the early and signed editions can be revisited, the cost of used volumes is not prohibitive, and some old treasures have been affordably reissued. At the very start I threw back any disturbing thoughts of turning what had started as a hobby into a profitable book. Among a good many fly-fishing friends and acquaintances, not one showed much interest in the history of the sport. Indeed, when I brought up some facet, the subject either got changed or the victim contrived to escape. I concluded that it's a good thing to be unburdened of expectations. The exploration and writing are fun, and if a few kindred spirits enjoy the journey as I have, that's enough.

In the course of my research, the backgrounds of fly fishing's pioneers, their motivations, the signal events in their lives, how they influenced one another, and even what it might have been like to have known them became as interesting as their achievements. And there were naturally societal and environmental spin-off issues such as the consequences of pollution and urban development over time, the privatization of prime fisheries (there's a caste issue for you), stocking practices and when and how "catch and release" came into being. Along the way I met an artist, a banker, several businessmen, a broker, editor, engraver, entrepreneur, inventor, lawyer, merchant, military man, physician, professional fisherman, riverkeeper, salesman, society figure, statesman, tax accountant, tackle maker and a tool-and-die man. Some were highly educated; others only completed grade school. There were blue bloods and blue collars, heads of large families, and a cadre of bachelors. Many lived to a ripe old age, others died young. As to temperament, the Hail Fellows Well Met contrasted with the less approachable, testy types. It would appear that the fly-fishing genes are not predictive of much about our lives. There was one exception though, a profile that began with a commitment to observe the details, large and small, that came with each day spent on the stream whether involving trout, birds, insects, other anglers, tackle, weather and water, and many other "factors." Our heroes were curious. They *wondered*. Hypotheses and experimentation were bound to follow, some as simple as trial and error, others quite sophisticated. They were *studious*. Surely you and I fish for enjoyment and relaxation, having finished our studies and passed our exams. For these anglers there was no distinction.

Hail Britannia and Dame Juliana

*I*t's a curious and somewhat convenient fact that the English and their neighbors from Scotland's southern highlands were largely responsible for the early development of fly-fishing, dating back to the fifteenth century. True scholars trace the earliest, cobwebby beginnings to China, the Macedonians, and the Romans, several thousand years BC, and it's said that the artificial fly was also being angled in Europe early on. Nonetheless, before the guys at the tackle shop can escape, I annoy them with the fact that the first sport-fishing book, *Treatise of Fishing with an Angle*, was supposedly written by an English nun, the legendary Dame Juliana Berners. There is an understanding, if not a covenant, among writers who intend to delve into fly-fishing history that they must acknowledge the good Dame's accomplishment up front. Actually reading the *Treatyse* is another matter. It's the archaic English. After a few sentences you start to wiggle in your chair. Fortunately, however the book has been "modernized" in John McDonald's *Quill Gordon* (1972). It begins philosophically with the assertion that "good and honorable recreations are the cause of man's fair old age and long life." Further, we are told, compared with hunting, hawking, and fowling, fishing has the most to recommend it. The *Treatise* contains a number of very proper, nun-like admonitions such as following the Golden Rule with respect to other anglers and warnings against trespassing or poaching. There are hands-on instructions for making a rod, a line, coloring a line, and fashioning hooks. We are warned of the "Twelve Impediments to Success." These still make sense and have to do with using proper tackle, the influences of season, of time of day, weather, water conditions, and so forth. The *Treatise* treats of many species of fish and of different baits and there are also recipes for a dozen trout patterns tied to resemble various insects. McDonald's book contains wonderful colored reproductions of these flies by the well known artist-angler John Atherton. Facsimiles were fashioned, working from Juliana's original descriptions, using the same materials insofar as possible, and carefully reproduced as to coloration. It's amazing. Although not intended to float, many closely resemble today's "no-hackle" patterns, upright wings tied with feather tips or bundles of feather barbs. Dubbed bodies are slender and tapered. Three pattern names include *dun*, consistent with an imitation of a mayfly subimago with dull-colored (dun) wings. Another label, Drake, is indicative of a mayfly imitation. The Shell Fly is suggestive of a caddis, and there are terrestrials such as the Wasp Fly. Tied in a different mode, a classic "soft-hackle" appears much as today. There's even a decent representation of a Woolly Worm!

A Problem

Historian McDonald devotes a lengthy chapter to a detailed examination of Dame Juliana's authorship. He concludes "that as author of the *Treatise of Fishing with an Angle*, Dame Juliana is a myth." In brief, McDonald could not find authentication of her involvement. Meanwhile there are cogent arguments that make her an unlikely candidate. Although the dates of Dame Juliana's birth and death are uncertain, it's accepted that she lived during the mid-fifteenth century whereas it is believed that the book was written between 1406 and 1420, long before the first printing in 1496. There is no question that Juliana became the prioress at Sopwell, close to the Abbey of St. Albans in Hertfordshire near London. She is known to have come from a noble family, and she was well educated, highly intelligent and possessed of great beauty. McDonald points out that while there is evidence that she produced at least a part of a book on hunting, the writing style is quite unlike that displayed in the *Treatise*. Another objection to her authorship of the *Treatise* hinges upon the idea that her cloistered, spiritual life would have been incompatible with the rough-and-ready sports of the day. Some nonetheless rose to Juliana's cause, arguing that her family background was such that, when younger and before taking her vows, she might easily have been invited to court and there participated in sporting activities. They suggest that she later used these experiences to her advantage. A sixteenth-century historian wrote, "Entertaining her mind with these honorable delights, and training her body she so arranged the whole course of her life that she either prudently might avoid illicit desires or vigorously overcome them. She was a Minerva in her studies and a Diana in hunting, lest by pursuing leisure she might be involved in the charms of Venus" (Mcdonald, *Quill Gordon*, 1972). (When I shared these sentences with my wife she remarked sourly, "A man must have written that.") George Pulman attributes the following rather earthy quotation to Fitzgibbon (unknown): "Miss Juliana herself must have been a lady of powerful 'thews and sinews' not very much macerated by prayer and fasting, prioress of a nunnery though she was" (*Vade Mecum of Fly-Fishing for Trout*, 1851).

The author of the *Treatise*, then, remains anonymous. McDonald suggested that he or she must have been a person of some education, perhaps a teacher, lawyer, or physician. The considerable body of information therein contained could only have been compiled by a number of observant anglers over a considerable period of time. It's believed that earlier writings must have existed, but historians have not found them. At any rate, the *Treatise* was anything but a tattered relic discovered in some dusty archive. Eventually appended to a separate book on hunting, hawking, and heraldry, *The Boke of Saint Albans*, it went through sixteen printings during the next hundred years becoming a "must read" for the serious sportsman. We are indebted to the good Dame, whether legend or myth, for providing a picture of the state of the art coupled to a time in history. More precisely, we learn that surprisingly lifelike representations of both adult and larval insects, forefathers of the dry fly and nymph, were being fished. The conceptual basis for the duel that was to follow some four hundred years later was at hand.

ALL ENGLAND, DIVIDED INTO THREE PARTS

*T*aste, in the sense of social status, was basic to the separation of fly fishermen into dry and wet camps in England and to an extent in America. Matters geographical and geological have both played underlying roles. This brief chapter, to set the stage for some upcoming differences in practice and opinion, describes the geographical and resulting cultural differences among British landscapes. From the angler's point of view England can be informally divided into three regions, the South, essentially south of London, the North, including the border with Scotland known as The Borders, and, appropriately, the Midlands in between. Regional viewpoints antedated the appearance of the dry fly, although by the latter part of the nineteenth century, "dry v. wet" had become an issue.

THE SOUTH

Logically, development of the dry fly should have taken place where regional streams were most conducive, that is clear, smooth, gentle of current, and rich in insect life. Resident trout would be expected to grow fast and fat. There should be natural protection from the ravages of flood, ice, extremes of temperature, and drought. This is the description of a chalk stream. Whereas chalk streams are scattered about England, the greatest concentration of these holy waters is found some sixty miles southwest of London in Hampshire, Wiltshire, and a half-dozen neighboring counties. If the region had a capital, it would be the ancient cathedral city of Winchester. The river so important to the dry fly, the Itchen, literally runs through it. Not far, the many-braided Test, picturesque Avon, Kennet, and a host of other chalk streams, large and small, arise from "downs." It's an odd old name for treeless, chalky uplands where rain is filtered before emerging in the form of headwaters, mineral-rich springs. Water temperature and flow volume (descent of no more than six feet or so per mile) are kept quite equable, which is conducive to the growth of plants (watercress is grown commercially), insects, and fish, especially large, overfed, sophisticated brown trout and grayling.

That's the geology and geography. Next came the impractical notion of successfully "imitating" an insect upon which the trout were feeding, leading to the design and manipulation of delicate tackle, especially buoyant flies. All this would require the efforts of talented, observant, and committed anglers. And isn't it likely that these would be well-educated men sufficiently wealthy to spend, possibly frivolously, the required time and effort? The great city, not far distant by rail, supplied the necessary cadre of gentleman fly fishers. Halford called them the Southrens, and they made up the geographical and social parts of the equation. Money mattered, since by as early as the 1850s expensive stretches of river or "beats" running through irrigated "water meadows," or picturesque, wooded parks, had to be rented individually or by a syndicate. (Lowland or water meadows flanking the Test and Itchen and other chalk streams were "drowned" in early season, filing a network of ditches and channels to improve the farmland.) My beat on the many-braided Test featured the manicured lawn of a country estate sloping down to the bank on the far side where a picnic table was flanked by topiaries and a gazebo. Domesticated ducks and geese constantly criticized my casting skills, for back casts had to unfold into narrow, leafy gaps between tall trees on the undeveloped opposite bank, and wading was not permitted. Regardless, I almost expected to be asked to tea.

THE NORTH AND BORDERS REGIONS

A formal border exists between England and Scotland. The plural, capitalized Borders is a broader *area* designation. Early on, the Borders extended from the River Tweed in the southern Scottish highlands into

Northumberland, an English county. (Today the term more precisely indicates an association of thirty-two governmental council areas, all on the Scots side.) The rugged terrain little resembles the lush water meadows at Winchester and is much less populated. Hadrian's Wall once reached across the area, and the city of Carlisle was involved in a series of bloody conflicts between the Scots and English, the battlefields of William Wallace and Robert Bruce. I've read that animal husbandry was once a major means of support and that reiving, what we would call rustling, as from your neighbor's herd, was a highly developed art. The English folks in the south, descendants of the Vikings, were a feisty lot as well. If we assume that the local anglers were rather a rougher lot than the gentlemen at Winchester, you can imagine how differences might have developed not to mention invidious comparisons between London and Edinburgh. One would expect, then, that a Borders fishing great might well have been a Scot, vigorous or even aggressive in his single-minded pursuit of the sport. And that description fits W. C. Stewart rather well. Not to get ahead of the story, Stewart may have been the inventor of the nymph and the upstream presentation! His home river was the Tweed and its tributaries were the sites of his experiments and resulting conclusions. Best known as a salmon river today, the stretches of the Tweed I visited offered long, gentle reaches alternating with well-defined current systems replete with pulse-quickening riffled edges and glassy tailouts whereas the higher elevation streams tended toward the swift and rock-strewn. There is no real demarcation between England's northern fisheries as in Yorkshire and Lancashire and the Borders above. It's said that the trout were once very plentiful if perhaps smaller and more accommodating than those in the Londoner's water and eager for the wet fly or worm. "Fishing the water" was routine. Indeed, the North Country style was known (in the south) as "chuck and chance it." T. E. Pritt of West Yorkshire, one of several northern fly fishers of note, extended and furthered Stewart's concepts and in particular brought the importance of caddis to a wide audience of fly fishers.

THE MIDLANDS

Like a wide cummerbund stretching across England's belly, the Midlands comprise some fourteen counties and roughly a third of the country. The region reaches from the boundary with Wales on the west to Lincolnshire and the North Sea to the northeast. Midlands defends its right to be capitalized on several counts. Historically there is an approximate geographic correspondence to the Anglo-Saxon Kingdom of Mercia. Geologically the Midlands is rather farther removed from navigable water than most parts of England and is sufficiently distant from London commerce that the region has developed its own agrarian and industrial economics. The terrain is rather flat aside from the mountainous area near Wales and the southern extension of the Pennine Hills, the Peaks District in the northwest. As might have been anticipated, anglers of both the dry fly and wet fly persuasions met along the Midlands waters. In 1899 Sir Edward Grey wrote, "Roughly it may be said that the dry fly method possesses the South of England, while the wet fly is superior in the West and North of England, and in Scotland. In the Midlands and in part of Yorkshire there is a disputed territory where both are used, and where there may be a real competition between them" (*Fly Fishing*).

Some of the most significant figures in angling history were Midlanders. The River Dove, intimately associated with Sir Izaak Walton and his companion Charles Cotton, arises from springs in the Derbyshire high country of the Pennines. In the second part of *The Compleat Angler* Cotton attests to the pride Midland anglers took in their fisheries, tracing the Dove, Derwent, Wye, and other regional headwaters of the mighty River Trent, the major arterial drainage of the Midlands. Today websites depict an inviting array of placid chalk streams as well as swifter, riffled streams. ("Fishing tickets" are rather expensive, just as in the south.) During the next century the Bowlkers, from Shropshire in the far west, produced the long-lived *Art of Angling*, regarded as *the* resource for many decades. Charles Bowlker's later editions offered the first comprehensive account of the art of fly-fishing. Better known still, Alfred Ronalds, from Staffordshire adjacent to the Dove, used the River Blythe as a laboratory. Among other accomplishments Ronalds provided a fly fisher's entomology, accurate for the time, keyed to remarkable color illustrations of insects and corresponding imitations. It was hatch matching's first blush. While Walton and Cotton had significant London connections, all indications suggest that their fishing was confined to "home waters." The same was true for Ronalds and also the Bowlkers, who frequented the headwaters of the River Severn as it began its descent to the Channel. ⫷

CHAPTER TWO

GRAYLING TOO

*W*ell known to anglers in Dame Juliana's day, the grayling played a substantial role in the early history of
fly-fishing. In the *Compleat Angler* Walton tells us that
the fish was also known as the Umber or Flower-fish.
Ronalds included a chapter comparing fishing strategies for trout and
grayling, while Pritt found the slender fish worthy of an entire book.
Both Halford and Skues regarded the grayling as a regular player in any
day's fishing, dry fly or nymph. And so while this introductory chapter
may seem out of place, the fact remains that the grayling, stream mate
of *Salmo fario*, was pursued in most of the same water with many of the
same flies. Some considered the grayling a game fish nearly equal to
the brown trout.

ORIGINS AND BIOLOGY

Robert Behnke, famous salmonid taxonomist, remarked upon the genetic constancy of this fish, still just one genus and four species (*Trout
and Salmon of North America*, 2002). Two are Mongolian exotics, the
others the common European and Arctic species. Intolerant of salinity
or pollution, native populations were not found in coastal rivers and
only entered English waters through stocking. Schwiebert posits that
Norman monks may have seeded English rivers in the eleventh century
during the reign of William the Conqueror (*Trout*). I like that romantic

notion, although Behnke disagrees. He explains that grayling, native to the Rhine River, swam *across* England during the last glacial epoch when the British Channel was a landmass and the Rhine entered the Atlantic near the mouth of the Avon. Thus grayling were eventually native, if only to that drainage (*About Trout*, 2007). They were not found in Scotland, and interestingly, the Emerald Isle's waters have never been planted. By the 1800s grayling were plentiful in Midlands fisheries. They flourished best in cold, swift streams at the foot of mountains and took hold readily in the southern chalk streams. Today large populations of good-sized grayling in peak condition are the pride of rivers in Wales such as the Dee. Relevant to stocking in America, grayling are highly susceptible to environmental stresses and disappear when non-native trout are introduced. According to Behnke, grayling were plentiful in the Upper Madison in the 1870s (*About Trout*). Freshly caught, the grayling is said to smell of thyme, hence the genus *Thymallus* and the erroneous idea that the fish feed on the wild herb. They are primarily insectivores, but will also take small fish and fish eggs.

Some Pros

"Lady of the Stream," an attractive name for grayling, recognizes the grayling's great dorsal fin, waving gracefully like a sail. In England Grayling offered the important advantage of providing good sport in late summer into November and even on to Christmas when the browns were going out of season. Being early spring spawners, the grayling were ready then to feed again right along with the trout. As to numbers, in chalk streams with plenty of trout, there were likely to be plenty of grayling. We can guess at this because anglers such as Skues made note of each day's catch. Regarding size, a two-pound grayling was a "nice fish" and at three pounds, a rather special catch (*The Way of a Trout With a Fly*, 1921). Halford's first publication was a piece in the *Field* in 1885 entitled "October Days at Houghton." This was on the Test, where in five days he recorded twenty-six grayling weighing a total of almost thirty-nine pounds. He concluded: "Thus the best sport may be anticipated by the fisherman who determines to treat grayling fishing as a serious branch of the subject, who uses small flies and fine gut, who fishes dry over rising fish." And he warned that at first sight of the net a grayling of decent size will bore downstream with smashing power and speed (*Floating Flies and How to Dress Them*, 1993).

Some Cons

Cotton was not a fan of the grayling, which he called the "deadest hearted fish that swims" (*The Compleat Angler, part 2,* 1676). He was irritated by their stubborn tendency to tug and shake rather than putting up a good fight. Frank Sawyer, inventor of the Pheasant Tail Nymph, looked at the fish from a different perspective. A professional river keeper responsible for maintaining a sport fishery on the Avon, Sawyer noted that "Grayling are appreciated by some and disliked by the majority," implying that for the less adept angler, "the gullible grayling often saved the day" (*Nymphs and the Trout,* 1958). Sawyer complained as well that the greedy fish "take insects that should feed the trout." Shamefully, they'd even take a dragging fly. He found grayling to be quicker and more aggressive than trout, even driving them away during a hatch. More often their stomachs revealed a preponderance of shrimp, snails, and other bottom foods, while the perceptive trout preferred the dainty mayfly. Sawyer said that, rather than pairing off to spawn like a pair of wholesome trout, the wanton females spew eggs randomly over the streambed for the randy males to fertilize (*Keeper of the Stream,* 2005). Grayling laid fewer eggs than trout, but their fecundity was much higher; in certain streams they bred so successfully that the trout population suffered (Behnke, *About Trout*). As you'd imagine, there were arguments over the question of whether grayling should be protected. During the 1920s some clubs and syndicates with leased water recommended the slaughter of several hundred of the fish each day during the season! Frank Sawyer was of the same persuasion, finding the fish no challenge for his nymphs. "I fish to kill," he declared. "I could easily have killed a few hundred that afternoon, but I got sick of the monotony" (*Nymphing and the Trout*). He liked to find a pod feeding in the shoals. If he was careful he could pick them off from the edge, one at a time, and if one broke him off, there was a good chance he'd soon get his fly back (*Keeper of the Stream*). (Although our Rocky Mountain Whitefish is not a very close relative, it shares this mixed reputation, thrown on the bank by some anglers and prized as winter "nymphing" sport by others.) While fishing the Test, my gillie admonished me to shun the humble grayling, pointing out, as had Sawyer, that a grayling's rise "humps the surface" rather than leaving a ring and central dimple. It wasn't easy though. They far outnumbered the browns, and their rise forms were almost indistinguishable (to me) unless a dorsal fin or forked tail flashed above the surface. I stuck one anyway when the gillie wasn't looking.

STRATEGIES

A combination of anatomical properties linked to feeding behaviors helps explain why grayling were sometimes called "silly fish." They preferred to lie deep, several feet or more, often toward the center of a river, whereas feeding trout hung close to the surface and were much more bank feeders. Sawyer killed larger Avon grayling in deep water fishing his "Killer Bug," a nondescript creation with an underbody of fuse wire wrapped to form a shrimp-like bump beneath a pinkish yarn body. The grayling took his Pheasant Tail Nymph as well (*Nymphs and the Trout*). Halford and Sawyer both noted that when grayling rose to take a floating insect they turned steeply, nose almost straight up, as they attacked. According to theory, the grayling's field of vision is smaller than a trout's and, having to ascend from well below the surface, the fish must time and position its impact accurately lest the small mouth miss the insect. This is why grayling were thought to so often "come short" or "rise short," that is, the fish appears to have struck at the fly, yet a well-timed response fails to set the hook; we've all been there. Ronalds gave the grayling more credit: A grayling, he explains, is said to "rise short" when "he does not seize the bait voraciously and confidently; and this want of zeal is no doubt frequently occasioned by the imitation shown to him being too faint a resemblance of the real insect" (*The Fly-Fisher's Entomology*, 1990). Interestingly, Halford writes of prolonged rises when almost every grayling "came short." Then, only minutes later, nearly every fish began to "fasten"—to the hook. Although "leader shy," grayling were regarded as somewhat less spooky than trout, particularly when cast to repeatedly, and might eventually take a fly that had been refused over and over when it was said that the grayling had been "worried" into taking (Halford, *Dry Fly Fishing in Theory and Practice*). While the same patterns were effective for both browns and grayling, small flies were tied especially for grayling. They were also more likely to succumb to bright, "fancy flies." (The parachute-hackled Klinkhammer is currently *the* fly in Wales I'm told.) Stomach contents revealed that surface-feeding grayling were less picky about their diet than the browns and had a special fondness for terrestrials. Halford said, dangerously, that their proclivity to sample this and that showed a "sort of feminine curiosity" (*Dry Fly Fishing in Theory and Practice*). As to table fare, grayling were not necessarily considered inferior to trout prepared in the same way although Walton preferred the latter. You wonder how so many were disposed of during systematic slaughters.

I like Sawyer's summation: "By some, grayling are considered to be the lady of the chalk streams—perhaps so, but I think a very greedy lady and a flirt. It is true they are nicely shaped and good to look at. To my mind they are just fish to interest one in idle moments—playthings, in fact, which give freely all that is asked until one becomes satiated with the ease by which they can be duped and then has an urge to try deception on something with more experience. It's easy to be irritated by grayling" (*Keeper of the Stream*).

Walton, Cotton, and the Hut on the River Dove

*D*espite the fame of *The Compleat Angler*, Sir Izaak Walton was not an especially important figure in angling history. Walton's gentle prose and charming celebration of the joys of the country life were what made him popular. His connections with fly-fishing were indirect. Indeed, historians regard Walton primarily as a fisher of baits! Much of what he had to tell us was borrowed from the *Treatise*. *The Compleat Angler's* importance to fishing is due to additions by Walton's close friend Charles Cotton, which constitute the second part of the great work (1653). (I have John Major's Fifth Edition, 1921.) Cotton's entertaining dialogue between Piscator and Viator provides a clear picture of how flies were fished in the mid-seventeenth century.

Anglers of the day must have struggled with whippy rods spanning some eighteen feet, not to mention lines made of braided horsehair tied directly to the fly. Despite their length, the limber rods could hardly propel the wispy line any distance without a breeze. Cotton wrote: "Endeavor to have the wind evermore at your back." As a result, the sport was largely restricted to the blustery spring months. When you do the math, though, an angler was not at such a disadvantage given an extended rod plus a line of similar length billowing out nearly forty feet over the water. In discussing how different species should be pursued, Cotton divided the water into three levels, stating that trout were to be fished for "on top" and with flies while the middle and deepest levels were reserved for the despicable gray-ling and other rough fish, tempted with baits or minnows. On calm days Cotton sought cover, either by kneeling on the bank or keeping its edge or a bush between himself and a rising or otherwise active fish. By carefully positioning the long rod he could "dap" the fly near the quarry. He called this technique "dibbling" and recommended that "not an inch of your line be suffered to touch the water." We can understand why the lightest line would be a necessity. Instead of X tippet calibers, Cotton talks of "two or three hair lines," tapered and inconspicuous yet strong enough to capture a good trout, prop-erly struck and played. (Incidentally, Cotton is credited for the famil-iar phrase "fine and far off.") At the time, and for centuries to follow, the meaty Green-Drake and Stone-Fly provided the most productive sport of all, in this case, impaled upon small hooks. "These Drakes," Cotton reports, "are taken by the fish to that incredible degree, that, upon a calm day, you shall see the still-deeps continually all over circles by the fish rising, who will gorge themselves with those flies, till they purge again out of their gills; and the Trouts are at that time so lusty and strong that one of eight or ten inches long will then more struggle and tug, and more endanger your tackle, than one twice as big in winter." So long as their wings were kept dry, Cotton was able to hover these "quick flies" at the surface for fifteen minutes or more. He even liked to have two on the line. Assisted by a strong follow-ing breeze we can envision how the flies could be presented "touch and go" reminiscent of caddis or stoneflies. (When I was a youngster, huge *Pteronarcys* stoneflies, locally called Willow Flies, were fished live and to deadly effect on Colorado's Gunnison River.) Cotton also enjoyed great success with imitations (tied freehand) of the Drake and the Stonefly. Detailed instructions for Viator take up over three pages. Several suggestions are as sound today as when Viator observes that the dubbing material is "very black." Piscator responds: "It appears so in hand, but step to the door and hold it up betwixt your eye and the sun, and it will appear a shining red. . . . and be sure to make the body of your fly as slender as you can." A loyal citizen of the Midlands, Cotton ridicules bulky, heavily tied London-style flies as "very un-natural and shapeless." When Viator's fly is finally taken, his mentor admonishes, "This is a diminutive gentleman; e'en throw him in again, and let him grow till he be more worthy of your anger." That's not quite catch release, but a start at least. Cotton followed the *Treatise* approach of matching month to fly pattern. *Dun* and *Drake* suggest that a mayfly was being honored. Whirling Dun persisted as a standard pattern name well into the twentieth century. Similarly, among terrestrials, the familiar Cow-Dung Fly made its appearance. The origins of dry fly fishing fall into two parts, the concept and the execution. Cotton observed trout and grayling taking insects from the surface, and choosing two of the most succulent, he fished both the naturals and his imitations "on top" as best he could. As we'll see, identifying the "who and when" of putting the floating fly into practice as we know it today is not so clear.

COLLEAGUES

Shared with "Father Walton" (Cotton was thirty years Walton's junior), Cotton's angler's hut, hard by the banks of the River Dove, has many descendants. These comforting structures of no more than a room or two provided shelter and were well stocked with ale, tobacco, and a bit of food to be shared with one's fellows. Discussions no doubt included fishing strategies. As we shall see, it's likely that the beginnings of dry fly elitism arose through such hut-centered conversations in Halford's day.

It's only natural, after reading about a famous river of long ago, to wonder about its status today. That's hardly part of a "history," but why not? Turning to a variety of websites, I found that the Dove is pretty pricey, currently as much as 1,300 pounds for a week during the Mayfly (Green Drake) hatch, and this does not buy exclusive rights; you have to share with club members. Only the dry fly and wet fly, fished upstream, are permitted and there is no wading. However, the fishing is said to be amazing. The lower river boasts opportunities for far more affordable barbel and chub.

Walton and Cotton were both royalists in the time of Cromwell. This and their shared love of fishing served to bridge the gap between otherwise disparate lifestyles. Walton, a pious and respected London businessman, had many friends among the highest clergy of the Church of England, whereas Cotton, quite erudite and charming, traveled in London's high society. (Talking to Cotton might have been like matching wits with Ben Jonson or John Donne.) He inherited two substantial estates and although married several times was apparently something of a rake and has even been described as dissolute.

He enjoyed writing poems, once considered ribald. If you'd like to read more about Cotton's licentious lines, Arnold Gingrich has a lot more to say in *The Fishing in Print* (1974). He reminds us, though, that what would have been considered shocking four hundred years ago might today be nothing more than a bit of sly naughtiness. Walton was buried with honor at Winchester, in his ninetieth year, but poor Cotton fell on hard times and was forced to sell his properties. He died at fifty-seven and was buried in a chuch in Piccadilly (Andrew Herd, *The Fly*, 2003).

Bowlker Anticipates the Future

*T*he eminent historian John Waller Hills traced the progress of fly fishing along two somewhat separate pathways that at times were not traveled at the same pace. He called one avenue the technical, referring particularly to mechanical advances in the tools of the sport. The other represented concepts derived from the "mind of man," biological observation and intellectual insights that influenced tactic and technique. At the beginning of the eighteenth century, tackle remained much as it had in Cotton's day. "At its close," Hills writes, "everything that we have now was in use except the American split cane rod. Reels, lines, gut, flies, net, basket; all were there." The ferrule had replaced the splice as a means of joining rod segments, "barrel winder" brass reels with a click were popular, silkworm gut had replaced the horsehair leader, ring line guides facilitated casting and hazel had given away to hickory, lancewood, greenheart or bamboo, far superior materials for the rod maker. Up to the end of the eighteenth century, as Hills pointed out, progress was technical rather than intellectual, "a progress wrought by the tackle maker rather than by the writer or thinker" (*A History of Fly Fishing for Trout*). Richard Bowlker's *Art of Angling* (1747) proved an exception. Taken over by Richard's son Charles, the book continued through sixteen more editions, the last in 1854. Hills commented that this longevity was surpassed only by the *Treatise* and *The Compleat Angler*. The Bowlkers saw fishing, broadly defined, as such a rapidly evolving practice that even their own earlier editions were somewhat disparaged. As a consequence the book was consistently and vigorously expanded and updated after Charles's death in 1779. The introduction to my copy, the 1833 version, upholds the theme of aggressive pusuit of progress: "In order to render this new edition of *Bowlker's Art of Angling* more deserving of general approbation, it has been carefully corrected, improved, and greatly enlarged: in every part of the work these improvements will be recognized and duly appreciated by the judicious angler; but especially in that part which treats on Fly-fishing." We know nothing about the editors responsible for additions and corrections over the next eighty years. Apart from the author's name, that of the printer and the date of publication, there are no further acknowledgments in my 1833 edition.

THE BOWLKERS

I've found no information regarding the Bowlkers' livelihoods or other details of their lives. We know that they hailed from Shropshire, a county on the Welsh border in the far west Midlands known for sheep production and corresponding textile industry. The Welsh were an unruly lot in the eleventh century, so William the Conqueror established a line of fortifications along the border. One of the most formidable was built high on a hill overlooking the River Teme in the medieval village of Ludlow, where the Bowlkers resided. A number of sizable rivers are found in Shropshire. The Severn, England's longest, takes its origin in the Cambrian Mountians of Wales, picking up the Perry and Teme, and further downriver, the Avon, before entering the Channel near Gloucester. (The Teme, presumably the Bowlkers' home river, was an exceptional grayling fishery and yielded a record specimen.) Ludlow castle and the surrounding village remain a popular tourist attraction today.

ART OF ANGLING

The slender pocket-size book was written as a broad instructional guide for anglers of all sorts who pursued fish of all sorts. Bowlker's concluding words verify that intent: "We have now completed our undertaking and having led our Readers through a regular course of instruction, founded on experience, teaching the true art of making artificial and selecting natural baits with a plain and comprehensive account of the best mode so arranging all the necessary appendages of the art, as to secure the adventurous Fisherman the pleasures of his favorite amusement in all seasons, regularly as they succeed each other." *Art of Angling* was the best pure fishing book of the time, according to Hills, and "was still in use as a text-book in my boyhood though written considerably more than a century before I was born." Gingrich referred to the *Art of Angling* as "almost an institution" and likened the book's wide readership and comprehensive treatment of the sport to Ray Bergman's *Trout (The Fishing in Print)*. Bowlker took his sport seriously and expected the reader to do the same: "Never regard what bunglers and slovens tell you, but believe that neatness in your tackle, and a masterly hand in all your work are absolutely necessary. About ninety in a hundred fancy themselves anglers. About one in a hundred is an angler." This passage suggests that there have always been dilettantes, as does another: "The Fantastico takes a great many things and kills a few fish. The old angler takes a few things, and kills a great many fish." And lest the Fanastico become too proud, Bowlker notes that in early season under ideal conditions "one might almost train a monkey to catch a trout." The chapters on trout and fly fishing are decorated with still useful insights such as: "Nothing will so soon, or suddenly, rouse a sick fish as the sight of a man or a landing net" ("sick" refers to a tired fish). "The first fall of the fly, in fishing, is like the first sight of a bird in presenting a gun—always the best." Translation: "First cast please!" Fishing in New Zealand, I became terminally tired of hearing that well intended admonition over and again. I always tried.

Matters pertaining to health were also important to Bowlker, and none more critical than the avoidance of wet feet. Most of a page is given over to oiling boots. Overheating was also a health hazard. "In your excursion to or from fishing, should you overheat yourself with walking, avoid small liquors and water as you would poison; a glass of wine, brandy or rum, is more likely to promote cooling effects, without danger of taking cold." Après fishing perhaps, but on the few occasions when I've tried this advice *before* entering the stream it left me all wet—literally.

A Fisheries Biology

Nearly half the book deals with the creatures of interest to fishermen. Trout and salmon are featured. Bowlker strongly suggests that anglers fish a fly as a means of catching salmon. "There is scarcely any time unless when it thunders or when the water is thick with mud, but you may chance to tempt the Salmon to rise to an artificial fly." The eel, "in some degree connecting the fish and serpent tribes," receives almost as much attention. We learn, in some detail, how to "sniggle" for the nutritious eel albeit with a warning against excessive consumption: "Eating of eels is hurtful to the throat, so say physicians of no common note." Some twenty other fish are also described. Handsome line drawings of the Pike or "River Wolf," the timid Chub, Tench, and Bream among others are keyed to a brief account of their habits, how they might be best angled for, and with which bait. There are others, too, such as the impudent "Skegger Trout" of brackish waters and the "Gwiniad," whose name has a mystic, Welsh ring to it but which turns out to be a rather handsome member of the Whitefish family found in alpine lakes in northern Wales.

Tactics

On occasion, Bowlker blurred the distinction between bait and fly fishing. "Dibbing" with captured insects was pursued, just as it had been in Cotton's time. And there was the bastard "fly-bait" consisting of two caddis worms threaded onto the hook toward the bend with a pair of wings of Landrail feather behind the eye. (I know a very successful nymph fisherman who, when necessary, seasons his fly with small salmon egg.) When hatches appeared, however, the angler should of course relinquish his baits and other methods for the fly. The following are Bowlker's most often quoted words:

> Prefer dropping the fly just under a bush or hedge, or in an eddy, to the open river, because your line is then obscured from the light, and the largest fish generally monopolize the possession of such places, in order to find and devour the more flies and insects; and, also, to be near their places of security. If the spot is quite calm, watch the first good fish that rises, avail yourself immediately of the ripple that has been made by the fish himself, and drop in your fly a little *above* where he last rose.

We'll come back to this advice in a subsequent chapter since Hills believed it to be the first recommendation for an upstream presentation. We learn, too, that streams were already over-crowded: "Trout will frequently cease to rise well, even at the best of times, from being every day whipped at by anglers from the same bank." My plan, in this case, is to go to the opposite bank and to throw against (or rather under) the wind. A friend and I once caught two and twenty brace by this means, while a whole tribe of professed anglers, who were fishing from the windward side caught (as we afterwards heard) but three fish among them.

Throwing a Fly

Bowlker's instruction uses few words, but they are well placed, descriptive and easy to follow. The "spring of the wrist" to and fro, properly timed and with sufficient power, does the job. The following might have come from any modern "how-to": "If, after sending it back, you make the counterspring *a moment too soon*, you will hear a crack and whip *off your tail fly*, and if *a moment too late your line will fall in a slovenly manner*." (*Slovenly* is the perfect word!) The young man with powerful, supple wrists is much preferred to the "poor fellow who makes no attempt with energy." Indeed the latter would likely "remain all his life in the background." Bowlker (or whoever) must have been a fast-action sort of guy. Included is a nice description of the cross-body cast (rod over left shoulder) in the case of a strong side wind from the caster's "right" side.

Tackle

The rod, twelve feet, three inches and weighing fourteen ounces, should be neither top nor bottom heavy. The suggested reel: a multiplier with click and loaded with a silk line and gut strand of three yards. (The reel was to be mounted on top of the rod and above the handle.) A fly book of "Russia leather" to repel moths filled the bill as did a "common beaver hat to hook and keep flies on." Extra lengths of gut with flies attached should be wound around. Bowlker's list of thirty patterns notwithstanding, on the water he found that "two or three kinds of flies" were usually sufficient. As to a carrying, as many do, a huge book of flies, nearly as large as the family bible, for common Trout streams, "He [the maligned Piscator] may go to the river side with a book of this sort, or even twelve pounds of lead in his pockets."

THE WINNOWING

That Bowlker paid an uncommon amount of attention to stream-side insect life is underscored by his description of the five stages in the life cycle of the mayfly. It's thought that the Green Drake was his model. The newly hatched "worm-like" larvae enter a chrysalis stage (in a burrow) before becoming "nympha" and eventually, flies. Accordingly, he felt that every pattern should be designed to at least resemble a particular insect. This belief led to the excommunication of a whole glut of antiquated patterns that failed to satisfy this criterion. These rejects he replaced with a smaller number of more imitative and productive flies, although not always new ones. This cleansing took place early on, for by 1833 the reader is pleased to find a color plate depicting thirty useful patterns, a fait accompli. Hills noted that after the *Treatise*, and following Cotton, most authors slavishly copied the earlier lists without further thought. Regarding Bowlker's efforts he wrote, "And it was high time that they went; for the original derivation had long been forgotten, their very names were corrupted and had become meaningless counters, unrelated to the natural insects from which they were copied." Gingrich noted that the "confusion was confounded and compounded by the totally un-systematized hodgepodge of vernacular names for flies that had accumulated like barnacles on the main body of fly-fishing literature" (*The Fishing in Print*). Bowlker actually singled out the deportees, including some "old friends" from the *Treatise*, and as Hills said, pilloried them individually. John McDonald credits Bowlker with a publication (possibly from an earlier edition) gloriously titled *A Catalogue of Flies Seldom Found Useful to Fish With*. This was presumably a preface to Bowlker's new list (*Quill Gordon*). A number, 1 through 30, was attached to each pattern and in turn matched to a descriptive paragraph in the text that blended the natural insect with its "imitation." Typically included were the type of water, time of year, and even time of day when the best sport might be anticipated, all followed by the pattern recipe.

TYING

As mentioned, the 1833 edition was apparently the first to incorporate fly fishing in a prominent manner. The short chapter devoted to fly dressing is noteworthy, for, according to Hills, the modern style started with *The Art of Angling* and he himself learned from the book. The clarity of the step-by-step instruction could hardly be improved upon, so much so that the mind's eye pretty well suffices for illustrations. That the tying is "hand held" is understood, and, unencumbered with details about vise and bobbin, the text reads smoothly, both for hackled and winged patterns. Color was important to Bowlker; for instance, a stickler for precision, he blended silk and fur (seal's) of the appropriate shade when dubbing a body to match its living counterpart. Schwiebert praised Bowlker's simple, direct approach for its emphasis on neatness and attention to size and conformation, qualities familiar in the classic Catskill dry-fly style that was to follow (*Nymphs*).

A PLACE IN FLY FISHING HISTORY

In the fishing literature of the eighteenth century, Hills said, "No great names stand out. There are neither great masters of the rod nor great masters of the pen." Regardless, a strong argument can be made that *Bowlker's Art of Angling* qualifies as one of the essential works. It served as *the* instruction manual for the average man on the stream for over a hundred years. And Hills did praise Bowlker: "His great merit is that he gives old ideas a good shaking up and fishing a fresh outlook. He clears away a lot of lumber." Equally compelling, as we shall see, Alfred Ronalds built upon Bowlker's foundation when he produced *The Practical Angler's Entomology* (1836). On the other hand, it feels uncomfortable to decorate Bowlker with full honors when so many other contributors updated and re-shaped the work long after his death.

RONALDS BRINGS THE SCIENCE

A fly fisher's life is made all the happier and more fulfilling through any of a number of related interests or hobbies—spin-offs, as it were. A lot of us tie flies; some build rods or nets. There are photography, writing, competitive casting, and, for the saltwater flats angler, the pursuit of a veritable grail, the elusive Permit. Other amusements involve those faraway fisheries where exotic or outsize trophies beckon. As an eclectic flyfisherman, however, no one has matched Alfred Ronalds. The list of his interests and accomplishments includes lithographer, engraver, and painter, all skills applied in illustrating his book, *A Fly-Fisher's Entomology*. He studied the trout's senses of hearing, taste, and smell. Attempting to understand what and how a trout sees, he turned to physics. Ronalds was regarded as a highly qualified entomologist and finally, and least surprising, a professional tackle maker. Correspondingly, his contributions to fly fishing history and to our modern sport are fundamental in both scope and impact.

Ronalds was born into a sizable and gifted London family in 1802. His brother Francis is said to have invented the telegraph and was eventually knighted in recognition of this achievement. Other members of the extended family included talented naturalists and illustrators. *A Fly-Fisher's Entomology*, published in 1835, went through twelve editions, was reprinted eighty-five years later, and continued in active use far into the twentieth century! *The Compleat Angler* is longer-lived, but as a fly-fishing entomology, Ronalds's work has no equal. Naturally the taxonomy of that time has been substantially updated, but the reality of the insects of importance to the trout and angler is the game, not the names we apply. And Ronalds was a practical man. He assumed, correctly, that fly fishermen would most appreciate depictions of insects upon which the trout were feeding matched to an effective imitation, both shown in color and side by side. And whether or not the readers cared about the taxonomy, it was there for them, Order, Family, Genus and species, matched to the corresponding "angler's name" for each insect, mayfly, caddis, stonefly, midge, alder fly, and a whole menagerie of terrestrials. Andrew Herd makes a related point: "The *Entomology* sold by the shelf-load because it solved the problem which had baffled fly tyers for centuries: the lack of a standard fisherman's nomenclature for waterside insects, a deficiency which meant that anglers from widely separated waters faced an impossible communication problem, as it was more than likely that they called the same insects by completely different names" (*The Fly*).

The Artist

My copy of *A Fly-Fisher's Entomology*, re-issued in 1990 from the final 1921 printing, boasts a three-by-four inch plate on the front cover displaying three mayflies opposite three corresponding imitations in realistic color. The finely detailed wing venation and body segmentation is perfectly sharp, as are the bars in the feather wings of the artificials. Ronalds was the fly fisher's Audubon. These delicate reproductions rival modern color photographs. Given that Ernest Schwiebert's own illustrations also set a standard, his appraisal of Ronalds's talent is especially meaningful: "The graphic work is beautifully executed and the copper-plate lithography remains equal to most modern printing technology" (*Trout*).

The Scientist

Ronalds is consistently described as the first scientific angler. In his late twenties, when he wrote the book, Ronalds was living in Staffordshire in the west Midlands, the River Blythe close by his home. His skills as an artist and craftsman notwithstanding, Ronalds was first a biologist whose studies focused on his home river, a laboratory for his experiments.

I built a little fishing hut of heath, overhanging a part of the river Blythe. Its form was octagonal, and it had three windows, which being situated only four feet and a half above the water, allowed a very close view of it. The curtains of the windows were provided with peepholes, so that the fish could not see his observer, and a bank was thrown up in order to prevent a person approaching the entrance of the hut from alarming the fish.

The book begins with a physical description of the brown trout and its habits. A detailed, keyed, and annotated diagram of a winding stream shows us where trout are most likely to hold because of depth, current, bank, and streambed structure. A first primer in water reading!

An experiment:

In order that we might be enabled to ascertain the truth of a common assertion that fish can hear voices in conversation on the banks of a stream, my friend, the Rev. Brown and myself, selected for close observation a Trout poised about six inches deep in the water, whilst a third gentleman, who was situated behind the fishing-house, diametrically opposite to the side where the fish was, fired off one barrel of his gun. The possibility of the flash being seen by the fish was thus wholly prevented and the report produced not the slightest apparent effect upon him. It is possible that fishes may be in some manner affected by vibrations communicated to their element, either directly, or by the intervention of aerial pulsations, it does not seem to be clearly proved that they possess any organ appropriated exclusively to the purpose of hearing.

(The lateral line sense organ was not yet understood.)

This is followed by the first description of the trout's famous "window," destined to be a critical factor in the theories and reasoning of many authors who followed. Once again the text is supported by a detailed, annotated diagram demonstrating the effect of the angle of incidence of a pencil of light intersecting the water's surface and the principles of reflection and refraction. Ronalds explains that the refraction of light rays enables a fish to look "round or over" the corner of a bank to spy an angler, provided that the fish is at a sufficient depth. He postulates that the image of an angler on the bank would be much distorted and indistinct and that the trout's eye may not be equipped for clear vision through air. Vincent Marinaro, whose observations in the mid-twentieth century so profoundly influenced the design of the dry fly, paid Ronalds the highest compliment by reproducing figure 1 from Ronalds's first chapter in his *The Ring of the Rise* (1976). "If Ronalds had lived and written his book in the dry-fly era, his genius most certainly would have been exercised to include surface flies in his studies of the effects of refraction instead of being limited to the appearance of the fisherman to the fish and the fish to the fisherman." Ronalds reminds us too that trout have a blind spot behind them and that a stream with a choppy surface obscures the trout's vision and can be approached boldly, observations that remain basic to the art of stalking.

> Regarding the sense of taste:

> I once threw upon the water, from my hut (by blowing them through a tin tube), successively, ten dead house-flies, towards a trout known to me, all of which he took. Thirty more with *Cayenne pepper and mustard* plastered on the least conspicuous parts of them, were then administered in the same manner. These he also seized, twenty of them at the instant they touched the water, and not allowing time for the dressing to be dispersed. The next morning the experiment was repeated when the same fish probably remembered the previous day's repast, and seemed to enjoy them heartily. I concluded that if the animal has taste, his palate is not particularly sensitive.

At the same time Ronalds found that trout were very leery about taking bees or wasps. I found this interesting in view of recent debates over whether trout feel pain. In any case he was certain that trout could quickly sense the consistency of whatever entered their mouths: "A fly should be made as soft as silk." (This suggestion was echoed by Skues and countless others since.)

RONALDS FOLLOWS BOWLKER'S LEAD

Theoretically a new and better imitation for each of the insects illustrated in *A Fly-Fisher's Entomology* might have been conceived, though that would have been a monumental task. Fortunately Bowlker had found a number of effective patterns, some even dating from the *Treatise*, that still lived on, and these became the core of Ronalds's catalogue, as he modestly called his book. The most important pairs (natural with imitation), forty-seven in all, are illustrated together on color plates in juxtaposition to the corresponding descriptive text. User friendly! Of these, at least eighteen came from Bowlker's list of thirty patterns! Ronalds's far advanced understanding of the relevant entomology came in part from the study of the stomach contents of trout and grayling put into a hair sieve, washed and separated for careful study. The practice was carried on by Halford, Skues, and, necessarily, every serious angler-entomologist since that time. Ronalds learned to recognize and copy various larvae and also came to understand the importance of terrestrial insects to include the blow fly, beetles, ants, wasps, the leafhopper, and three caterpillars, all depicted, along with an imitation. Ronalds covered the water *and* the waterfront. Of the aquatics, twenty were mayflies, four each stoneflies and caddis, with a couple of midges and the water-boatman to fill out the roster. Each of the forty-seven illustrated insect/imitation pairs received, on average, a half-page of text adjacent to the corresponding color plate. Very much as in the *Art of Angling*, he identifies the insect's origin, nymph or egg, when during the season it appears, and for what duration. The most likely time of day is often included. Whether the pattern is effective for both trout and grayling might be mentioned as well. (There is nothing to suggest that the grayling was considered an inferior prize.) Then comes the pattern recipe and finally hints for the tyer. The pattern labels in Ronalds's index most often refer to some combination of the color or size, also true of Bowlker. A few names relate to where the creature is found, such as "Cow Dung," and the rest are ambiguous, even dating back to Dame Juliana. Ronalds admitted that there was much more to be learned. For example, he was forced to lump together many lookalike species, say a group of mayflies, similar in size, coloration, and behavior, appearing at the same time of season. He realized that conventional ideas about certain mayflies must be faulty, such as the notion that the duns of a particular species hatched in the spring while the corresponding spinner was thought to be on the water all summer.

While Ronalds recognized that exact imitation was impossible, color was important to him as it had been to Bowlker, even subtle shades, and various dyestuffs were discussed at length. Organic materials included the outer sheath of the onion root, logwood, sumac, and fustic (a tropical mulberry). Copper salts were important among inorganic chemicals, as were potash, tin nitrate, alum, and hydrochloric acid. Arriving at the proper pH (degree of acidity) was crucial, and lacking a means of measurement, Ronalds recommended the tongue: "If it [the acid] burns your tongue much, it will burn the feathers a little."

Comparing the natural to its imitation seems an obvious approach, but no one before, or for many years thereafter, did the job effectively. In truth, first Bowlker and then Ronalds developed the style of modern treatments replete today with excellent color photographs, charts, graphs and the like.

A Puzzle

Ronalds fully understood the life cycle of the mayfly to include its "nymph," yet restricted his imitations to duns and spinners, the former tied with wings of realistic size, partially erect. Why this approach when the flies were fished wet as a brace? Andrew Herd offers a most interesting conjecture: ". . . critical analysis of his book implies that he fished his patterns as floating flies and must have intended others to do so, but the reality was that by the nineteenth century, Ronalds' patterns found themselves being fished more and more as wet flies and by the time of the tenth edition of the *Flyfisher's Entomology* was published, the change was complete" (*The Fly*). In Ronalds's comments on presentation he recommends that the flies alight gently and be left to drift for several seconds before the line tightens. As had Bowlker, he suggested casting to the rings of a rise as quickly as possible (the flies presumably still on the surface) or dropping them just upstream of the rise. I found nothing to further support the dry fly thesis. Fifteen years later G. P. R. Pulman offered a far clearer description of fishing the dry fly although suggesting many of the same patterns as had Ronalds (*Fly-Fishing for Trout*, 1851).

You marvel that, still a young man, Ronalds displayed such a diversity of knowledge and skills, all the while supporting a family. He was enjoying profits from his book but, for whatever reasons, moved the family to various locations in Wales where he was engaged in the fly-tying and tackle business. Following the death of his first wife, Ronalds emigrated to Australia, re-married, and worked as an engraver and copper-plate printer. An unprofitable foray into the gold fields followed in 1851. At the time of his death following a stroke, age fifty-eight, he was engaged in farming. Andrew Herd offers a much more detailed biography of this remarkable man (www.flyfishinghistory.com/biographies.htm).

A Good Fly is a Good Fly

What was the eventual fate of the genesis patterns from the *Treatise*? Originally copied from nature, in certain cases the design of the fly and its purpose (to represent a particular insect) remained much the same for centuries. As an example, Hills reviewed the life history of the Red Spinner. Walton listed it as the fourth of twelve flies "likely to betray and condemn all the Trouts in the river." His Ruddy-fly honored a mayfly, anticipated in early May "the body made of red wool wrapt about with black silk, and the feathers are the wings of the drake . . . with the feathers of the red capon" (*The Compleat Angler*). The same pattern later passed Bowlker's inspection, but now labeled *Red Spinner*. Ronalds thought the fly important and in 1835 wrote, "This is the name given to the Blue Dun, after it has cast off its olive brown coat. It now appears as a bright, red brown, and its wings, which were before rather opaque, are transparent." At the close of the century the pattern was one of Halford's staples for a summer evening's rise (*Floating Flies and How to Dress Them*, 1887). Writing in the 1920s, Hills comments that the same fly, by then called the Red Quill, could be found in any tackle shop. That's a span of five-hundred years! And the Red Spinner is illustrated in English author John Cawthorne's *The Fly-Fisherman's Entomological Pattern Book* (2000). Red game hackle, wisps for tails, a reddish, dubbed body ribbed with fine gold, and pale grey, cock-hackle-tip wings. The bird or beast contributing the wings and dubbing changed from time to time, but the "look of the fly on the water" had hardly budged.

The Shell-fly proves that caddis imitation also goes back to the dawning. This pattern was tied with "a grene wull & lappyd abowte wyth the herle of the pecoks tayle" and listed as Dame Juliana's choice for July. In his *Quill Gordon* John McDonald quite reasonably wondered if "shell" might not refer to a caddis case, and historians generally accept the Shell-fly as a caddis imitation. Bowlker happily passed the fly on to future generations as the "Granam" or "Green-tail." Ronalds called the pattern the Grannom, accepting it as the original Shell Fly. The prominent green egg sac, compatible with the modern genus *Brachycentrus*, suggests that July would be very late for the emergence of that insect. Nonetheless, Ronalds wrote, "it is a good fly all the summer months and into September." The English entomologist Martin Mosely eventually solved the riddle by identifying the insect as a member of the genus *Lepidostoma*, hatching in mid-summer rather than the early season (*Dry-Fly Fisherman's Entomology*, 1921). The Grannom, however Latinized, spanned the Atlantic, appearing in Mary Orvis Marbury's *Favorite Flies and Their Histories* (1892) and was listed synonymously as *Green-tail* and *Shell Fly*. Halford's version of the pattern is also illustrated. If we don't often see "Grannom" anymore, the basic pattern has done noble duty over the centuries when any of a variety of caddis with green egg sacs are out and about. ⋙≺

Stewart Faces Up

*S*tewart is a great Scots name. True to character, William C. Stewart attacked his angling with a Celtic fierceness and resolve, for he was, in part, a professional market fisherman, adept with bait or fly. In a photo reproduced in the 1996 edition of his *The Practical Angler*, first published in 1857, we see a bearded man of stern countenance, staring back with narrow-eyed intensity, meanwhile pouring himself a libation. Scotch? Why not! Historians regard Stewart as the most important of all the Scottish fly fishing clan and that the equal of Halford, Skues and a number of American icons. *The Practical Angler*, published in 1857, was his only book, yet his accomplishments were considerable:

- He introduced (arguably) the practice of fishing wet flies *upstream* (worms too) and proved the effectiveness of the method. This was a signal departure from the past and an admission that Cotton's flies necessarily followed the wind. McDonald described this contribution as the "First leap forward in fly-fishing since its origin" (*Quill Gordon*, 1972).
- He developed a unique "touch and go" style of presentation.
- He invented the "spider," the forerunner of today's "soft-hackles."
- He wrote the widely read book mentioned above, a classic primer that documented existing tackle and techniques for both fly and bait fishing.

- He provided an early description of fishing flies in lakes.
- He served as spokesman for and defender of the northern and Scotish Borders rivers and streams *and* those who fished them.

BIOGRAPHICAL

Richard Hunter provides some background in The Flyfisher's Classic Library edition. Stewart was born in 1832 in Edinburgh, where his father was a successful wholesale merchant in wines and spirits. Hunter believes that young William's angling career began during frequent visits to his grandparents in the Orkneys. Beyond the fact that he was educated in Edinburgh, where he may have studied law, it has proven difficult to trace his whereabouts and activities until he was in his twenties. Given that his detailed and comprehensive book appeared when Stewart was only twenty-five, we can guess that a great deal of time had been spent with rod and reel. My edition includes a half-dozen very positive reviews of *The Practical Angler* when it was first published. Some indicate that Stewart had already developed quite a reputation, and that his career as a pro-fessional fisherman was well established. The preface also lets the reader know that the book is written in dead earnest, unlike most books on angling. The dour Scot! In his classic *Fly-Fishing* (1899), Sir Edward Grey mentions a jealous gamekeeper who attributed the aggressive Stewart's huge catches to "twenty-four hours of creeping and crawling." "Angling, when once embarked in by any person possessed of a reasonable amount of soul and brains, becomes a passion, and like other passions will grow and feed upon the small-est possible amount of encouragement." (Anecdotally, neither William nor his brother Edward, a frequent fishing companion, ever married.) Stewart took great exception to Samuel Johnson's snide remark made about a century earlier: "Angling means a rod with fly at one end and a fool at the other." If the young Caledonian had not been long in school, he had considerable ability as a writer, perspicuous in his instructions and quite capable of engagingly descriptive prose, enlivened with wry bursts of humor. *The Practical Angler* is a treat to read, the text sprinkled with odes to the noble River Tweed and snippets of poetry: Burns, of course, and Lord Byron, too. There are quaint passages like this one by James Hogg:

"The wildered Ettrick wanders by, Loud murmering to the careless moon." During the off-season Stewart engaged in "commerce," possibly with his father and brother.

EVOLUTION OF THE UPSTREAM NYMPH

I'm always tempted to identify "firsts"—*the* angler who first achieved a breakthrough or developed a critical concept. For example, I would very much like to have awarded the prize for "*First* angler to fish a wet fly/nymph upstream" to Stewart. It would have made me feel smarter than I am. But the true historians who dig through the nitty and gritty can always find an earlier and less-known candidate and recognize too that no one really knows, for the prize often goes to the angler who first put whatever down in words. The pen is mightier than the rod. Nonetheless, Stewart's role in this regard is central to his place in history. Andrew Herd documents that acceptance of the upstream method came grudgingly and slowly, as was also true of acceptance of shorter, stiffer rods, ideal for Stewart's technique (*The Fly*, 2003). Even more controversial is the suggestion that Stewart may have been the first nymph fisherman. What would it have taken to earn this designation? First, he (or she) must have recognized that aquatic larvae and caddis pupae constitute an important food form. Accordingly imitations, or look-alikes anyway, should have been intentionally created. They would then have been presented in a manner miming the activities of the naturals. Finally, trout feeding upon "nymphs" should have accepted the artificials with at least a certain grace and enthusiasm. Stewart was thoroughly acquainted with Ronalds's book and had a solid, practical knowledge of insect life. Earlier in the season he focused upon the "Mayfly," actually an inch-long stonefly, apparently given the erroneous name in the Borders districts by the locals since it appeared in May. He fished both the adult and larva or "venomous-looking creeper," *but as bait.* I've found no indication that an attempt was made to imitate the larvae. Rather Stewart informs the reader where to find the creatures and to be sure to collect plenty. His experience had been that commercial imitations of one insect or another were seldom as productive as old-time general patterns.

A Point of Confusion

It's all about Mayfly—capitalized. In England from Ronalds's time until
the present the word has been reserved, with honor, for the Green
Drake. The custom is as British as afternoon tea. What about all the
other Ephemeroptera? As in Ronalds they are frequently known by the
name of the patterns anglers had attached to them or some descriptive
term such as the July Dun or Orange Dun. Thus when you see phrases
like "during the Mayfly time" or the "annual Mayfly madness," they
refer to the Drake emergence of three weeks, more or less, at the end of
May extending into June. John Cawthorne produced a useful reference
for English fly-fishers in which Chapter 1 is entitled "Mayfly," Chapter 9
"*Ephemeroptera* (Nymphs)," and Chapter 10 "*Ephemeroptera* (Adult)"
(*The Fly Fisherman's Entomological Pattern Book*, 2000). To make matters worse, the Green Drake spinner was long known as the Spent
Gnat. The reasons have been lost in time. (Cawthorne no longer uses
that term.) It gets worse. According to John McDonald, "'Mayfly' has
been applied to any of five distinct species belonging to two different
genera and two different families" (*Quill Gordon*). Cawthorne states
that today there are two Mayfly species, *E. danica* and *E. vulgata*.

Spiders and Winged-Wet Flies

Stewart leveled two criticisms at commercially tied flies. Practical
anglers do not put on one-third of the feathers that town-made ones
have. Secondly, the thread of gut on which the fly is dressed is of
more importance than the fly itself. He is referring to thickness (the
less the better) and transparency (superiority of clear gut). Spiders,
distinctly his invention, had no wings or tail. Dubbing, if any, was
thin and confined to the front half of the hook. Rather than simply
tying in the feather butt and winding on as most of us do, he first
spun the anchored feather *around the silk thread* and wound both on
together, covering only the front portion of the hook shank.

> With the hook held secure in the left hand take the thread, lay
> it along the centre of the inside of the feather, and with the fore-
> finger and thumb of your right hand twirl them round together
> till the feather is rolled around the thread; and in this state wrap
> in around the hook, taking care that a sufficient number of the

feathers stick out to represent the legs. The hook is better concealed, and if made of sufficiently soft material, the water agitates the feathers, and gives them a life-like appearance, which has a wonderful effect, and is of itself a sufficient reason for trout preferring them. . . .

Is this "nymph speak"? Not in the sense of intentional imitation, for his illustration shows quite a mat of fibers about the length of the shank. It's true that many later writers also attributed the success of their patterns to the lifelike mobility of the pliable hackle fibers resembling swimming or squirming larvae or caddis pupae. Schwiebert and several others have suggested that spiders made a good representation of caddis pupae, and Stewart does describe success with Spiders during summer evenings and nights when fish were rising, *but never mentions caddis*. He paid attention to the size of prevalent insects, smaller flies in summer and larger at night (remember, he was a 24–7 sort of angler), and was well aware that trout could exhibit sharp preferences for color. However this was regarded as a sort of silly caprice. Their whims could be satisfied with flies tied in just four colors, dun, black, brown and red. Stewart sometimes fished as many as four flies of different hue on his leader, taking note of their individual scores and adjusting his team accordingly. His favorite feathers came from the neck of a cock starling or the wing of a landrail. Unfortunately, the spiders were rather delicate and would unravel after a taking a dozen or so trout, a quantity most of us would not complain about. Partly on this basis Stewart also tied three lightly dressed wet flies with slender backswept wings in the same colors. These were more durable, and at times the trout took them as readily as the spiders. Schwiebert suggested that these patterns might suggest drowned or egg-laying adults and called the Blue-winged Hare's Ear "a superb imitation of several emerging mayflies" (*Nymphs*, Vol. I, 2007). Skues agreed that there was a resemblance between the thin wings of these patterns and the crumpled, compacted wings of an emerging insect: "during the rise, they are often, no doubt taken, for what they purport to represent—namely, the subimago hatching or hatched out" (*The Way of a Trout With a Fly*, 1921). Once again, whether Stewart intended to suggest an emerging mayfly or caddis, he never told us.

PRACTICAL AS ADVERTISED

Convinced that attempts at close imitation (of the adult insects) were futile, even foolish, Stewart wrote rather sarcastically: "And we believe the angler who has a different fly for every day in the season will kill *nearly as many* trout as the angler who adheres to three or four varieties the whole season through." The great point then, in fly-dressing, is to make the artificial fly resemble the natural insect in shape, and the great characteristic of all river insects is extreme lightness and neatness of form. Stewart watched for ways to increase his success, constantly devising experiments to test his hypotheses. For example does the tail or "stretcher" fly in a cast of three or four flies enjoy an advantage? He postulated that perhaps, in addition to the protruding dropper strand, the trout sensed the close-by leader stem? Stewart did the experiment, fishing multiple copies of a successful artificial simultaneously. Sure enough, the tail fly won out. Are trout sometimes put off by the hook? Take two identical flies. Leave one intact and clip the hook at the bend on the other. Will more fish attempt to take the neutered fly? They did.

FISHING UPSTREAM, QUICK AND SHOAL

It was not in Stewart's nature to tell us how he determined that short, upstream casts, his fly kept close to the surface, were so effective. He simply gave instructions as to how to proceed: "The great error of fly-fishing, as usually practiced, and as recommended to be practiced by books, is that the angler fishes down stream, whereas he should fish up. We believe that we are not beyond the mark in stating that ninety-nine anglers out of a hundred fish down with the artificial fly; they never think of fishing in any other way, and never dream of attributing their want of success to it." It's possible that Stewart learned the practice from James Baillie, a countryman whom he greatly admired—"the ablest fly-fisher in Scotland." (I've no information regarding Baillie.) Stewart goes on: "We believe that a mere novice would, in a large water, catch more trout by fishing down than up, because the latter *requires more nicety* in casting. But to attain anything like the eminence in fly-fishing, the angler *must* fish up, and all beginners should *persevere* in it, even though they meet with little success at first, and they will be amply rewarded for their trouble." Others were surely dabbling with the upstream wet fly, but it was Stewart who codified the tactic.

Stewart's technique was precise and neatly explained. He reasoned that trout face upstream and are thus more safely approachable from below. Cast upstream, the hook is drawn into the trout's jaw when setting rather than being pulled away. When a fish is hooked it can be quickly netted from below, preventing it from disturbing the water above. And finally, the angler can better adapt the motions of his flies to those of the natural insect. Here he is getting at the observation that downstream fishermen "played" their flies a great deal, skipped, danced and raced them across the current, behaviors Stewart thought quite unnatural as compared with a downstream drift. Stewart found that trout actively feeding on the surface during a hatch would often accept a wet fly carefully presented a foot or so above their lie, and he continually emphasized "fly before line." This he accomplished with a forceful forward cast, rod stopped high, and the point raised when the fly is "just about alighting." (It sounds like our "tuck cast.") And he strongly advised that the line never be longer than necessary. His reasons included easier line management, better accuracy at close range, less danger of drag with as little line on the water as possible, and a better chance of setting the hook. The take was either immediate or came after a few seconds of natural drift, the moment the fly alights, being the most deadly of the whole cast. When simulating a hatch he'd flick the flies onto the surface just long enough to let the trout see them before lifting off. "After your flies alight, allow them to float gently downstream for a yard or two, taking care that neither they nor the line ripple the surface. There is no occasion for keeping your flies on the surface; they will be quite as attractive a few inches under." Silk lines were by then the norm and he denied the possibility of fluttering a fly on the surface effectively as Cotton had with long rods and feather-light hair lines. His touch-and-go technique was never more effective than when mayflies or stoneflies were thrashing about on the surface. If rising fish were not about, Stewart fished aggressively, quickly covering every seam and probing any nook or cranny that held promise, casting constantly and picking up after drifts of a yard or two. If feasible, a direct upstream presentation was avoided lest the fish be "lined." Stewart had already experimented with other techniques that were to be regarded as novel in the century to follow, such as repeated presentations to a particularly inviting target in an effort to simulate a hatch. (A light, stiff rod of about nine feet was preferred.) When river currents were too swift and deep to wade against, he sensibly progressed downstream, but still with across and upstream casts and abbreviated drifts. Two of Stewart's favorite streams were the Whiteadder and the smaller Blackadder, both in the watershed of the Tweed. They are described as typical freestone "pocket water" fisheries ideal for Stewart's style. (I've not seen them.)

Well, that's the evidence supporting Stewart as the pioneer of upstream nymphing. If collecting and copying the larva or pupa of a select insect of interest to the trout is a necessary qualification, he fell short. Nevertheless, Gingrich wrote in *The Fishing in Print* (1974), "Purism was fathered by Stewart's upstream technique on the body of Ronald's knowledge." Here he refers to the dry fly purist casting a dedicated imitation upstream to a rising fish. It's clear that Stewart's preference had nothing to do with the idea that fishing upstream was more ethical or gentlemanly. As Schwiebert says, it simply added more weight to his bag. Schwiebert concludes, "The later investigations of Skues would not have progressed at such a felicitous and successful pace without the paradigm of Stewart and his subtle border-country patterns of sparsely dressed wet flies" (*Trout*, 1978).

The Sparks Fly

Stewart had a rather powerful adversary with the unlikely name Francis Francis. The editor of a leading sporting journal and influential spokesman for the gentlemen at Winchester, Francis incurred Stewart's wrath by suggesting that noble English (chalk stream) trout only succumb to "exact" imitations of whatever naturals might be on the water, thus necessitating a large assortment of patterns, whereas a few flies would suffice for the Borders angler who counted his catch of puny and primitive fish by the mere brace. (This sentiment was offered before the dry fly had become a central issue.)

Stewart replied, "Fishing in preserved water loses a great part of its pleasure. We like to be free to seek trout where we like, and take them where we can; and as there is more merit, there is more pleasure, in filling a basket where all anglers, high and low, rich and poor, are free to do the same, than in a river, fished only by a favoured few." Our liberal Scot was just warming up. Taking a dirk from his sock:

"And on behalf of Scotch anglers we repudiate with scorn the bare idea that it requires less skill to catch a Scotch trout than an English one, or that the former in any way receives an inferior education as regards flies, etc. to his English brother. In fact, we believe that in the before-mentioned streams the education of the inhabitants is as superior to that of the inhabitants of English streams as the education of the people of the one country is admitted to be that of the other." Take that, you Southrens!

Elsewhere Stewart adds a caveat to the effect that the fine trout he describes are found in heavily fished rivers such as the Tweed and Tay. He did discourage angling in the headwaters rills where smaller fish are not worthy opponents and "careless habits may be formed from fishing too much for them."

The Competitive Angler

The measure of an angler's degree of prowess deserves comment. A respectable day's catch apparently should have weighed no less than twelve pounds, whether caught with flies, worms, or minnows. This was not a legal limit, though. On a good day that poundage might easily have been doubled or tripled! The trout were carried in a cloth bag with a long strap slung over one shoulder, the bag resting against the opposite thigh. Imagine lugging around a catch of twenty-five pounds near day's end, the ache in your shoulder, a stiff neck to follow, and what about problems with balance? Stewart mentions that the serious angler (probably a professional) should bring a bag carrier or trout caddy along. The effects of increased fishing pressure and illegal netting were making the fishing much more difficult. Stewart had this to say about the busy streams: "So far from looking upon the increase of anglers with alarm, it ought to be regarded with satisfaction, the more trout are fished for, the more wary they become; the more wary they are, the more skill is required on the angler's part . . . angling is much better sport now than it was fifty years ago." I was interested to learn that "fishing competitions" were already in vogue. Hunter indicates that of six such, Stewart won three and tied in a fourth. For Stewart, it wasn't *just* a business!

Stewart took some shots at the foibles of certain of his fellows. "We have never met a bad angler that had not a good excuse. Anglers have an extraordinary knack of raising, hooking and playing, but losing large trout. The trout, once escaped, there is ample scope for the imagination to conjecture its probable size." A famous angler bragged of his day's catch while returning from the river by train with Stewart and friends. Suspecting fraud, they secretly opened the expert's bag to reveal fish so small Stewart called them "troutkins." He said "people who do not put a padlock on their baskets ought to put one on their mouths." An avid hunter, Stewart was intensely competitive and is said to have gone full bore into any activity that interested him.

OF LOCHS

A chapter in *The Practical Angler* is devoted to fishing lochs. It was not Stewart's preference since he felt it required less skill. "There is the tiresome monotony in fishing continually in still water, where the angler must ply his lure unremittingly in the same manner— many of our lochs are situated amidst the finest and grandest scenery in the country, the angler must indeed be destitute of taste if he can find no enjoyment in it." Stewart apparently fished Loch Leven with considerable success using a limited number of flies, slim of body and wing, and in varied, sometimes bright, colors. And his presentation was much the same as in streams, multiple, short casts, the flies near the surface. Of interest is Stewart's description of the monstrous *Salmo ferox* trout, genetically a brown trout, but highly predatory. A specimen weighing just under forty pounds had recently been caught on a fly. Behnke confirms that certain wise old trout move to the depths of these lochs where they feed upon char, growing ever larger and ultimately living longer than their brethren (*Trout and Salmon of North America*, 2002).

OF WORMS

Stewart fished worms upstream in much the same manner as flies and with the same attention to detail. He regarded the practice as something of an art form. A fourteen-foot rod allowed him to deliver the bait gently lest it drop off and with his brother Edward, proved that a three-hook rig was superior to the standard bait hook. As summer came on the worm had taken the place of the stonefly and Green Drake. This was clear water sport, too. Trout would even take the worm during a thunderstorm, "that dread of the fly-fisher." Stewart reports a catch of over twenty-two pounds, the excitement ending when the water finally came thick and muddy. On another occasion he caught thirty-six fish in a single hour—that during a mayfly hatch! It was recommended that no fewer than 200 worms be packed for a day on the stream. And he is not worthy of the name of angler who cannot, in any day in the month (July in particular) when the water is clear, kill from fifteen to twenty pounds weight of trout in any county in the south of Scotland. Stewart fished the wriggler only when necessary, though. Anglers who used them all season long were looked upon with disdain.

Of Minnows

The minnow received an entire chapter. "No method of trout-fishing exercises a more lively influence over the angler's hopes and fears, or requires the exercise of so much presence of mind as trouting with the minnow or parr-tail. . . ." The live minnow excelled in the removal of some of the largest trout of all in early season and tested the angler's skills to the fullest in low, clear water. At times, though, capturing the minnows was the more difficult challenge. We are told how to rig minnows to make them flash and spin, much like a modern lure and the benefits of a "dragger" or trailing hook. The ideal minnow was found to measure just one and three-quarter inches, cast with a sixteen-foot double-handed rod.

Of Salmon

Stewart's home river was the Tweed, even now much touted for its salmon runs, as were many other Borders rivers. He was considering a book on the subject just before his death. However, the trout season ran from early spring into fall whereas the salmon run was relatively brief, so in order to remain productive a market fisherman would have to have concentrated on the trout. Stewart even complained that salmon par consumed so much food there was little left for the trout. It may be too that the laws at that time restricted the salmon catch. Stewart derided the practice of taking salmon illegally or using the eggs for bait. (Numbers of salmon were killed during the closed season, in large part for their roe.)

Fisheries Conservation and Environmental Concerns

We've understood that Stewart viewed his fishing as "sport," and there is no mention in his book of any aspect of his commercial enterprise. It's interesting, nonetheless, that certain others were concerned about the killing of large numbers of trout, regarding the practice as "butchery." Stewart disagreed, indicating that it was his intention to show how the greatest weight of trout can be captured in a given time. "We cannot see the justification of an opinion that considers the capture of a certain number of trout sport, and of twice that number—taken by the same means—butchery. If the sport of angling lies in the capture of fish, it seems evident that the more fish the better sport." (And the more profit.) Fishing as a means of gaining a livelihood *and* as a sport seems contradictory, but in times when fisheries were overflowing with their bounty, this was the rule.

Stewart was acutely aware of the impact of industrialization along the major waterways such as the Tweed and Gala. Expanding agriculture played a related role in that wool was required to make yarn to be dyed and woven into tweed. Sheep, overgrazing the high hill country altered runoff. Heavy storms caused short periods of severe flooding rather than raising water levels less dramatically for longer periods. Floods shifted the gravel and carried off the spawn and also the eggs of aquatic insects. Then there were "bleach fields" which "send their dyes and other refuse straight into the streams, causing sad havoc." A segment of the Tweed was described in the *Border Advertiser* as "an unseemly ditch, full of the blackest and most obnoxious dyes." Predictably, a commercial or "conservative" group criticized the editors. Stewart clearly intended his book to make an environmental plea, closing with: "Shall the battlements of Abbotsford be reflected from a monster gutter, all stains and stench?" (Home of Sir Walter Scott, Abbotsford was and is still a shrine.)

A Pox

William Stewart lived just sixteen years after the publication of his book, succumbing to smallpox when he was just forty. His death came over seventy years after the prevention of smallpox by inoculation with cowpox had been demonstrated, but the disease was still far from con-trolled. (In America it remained a threat even in the 1920s.) There were then, even as now, considerable difficulties in surmounting the social and governmental barriers that stood in the way of universal immunization and public health. ◄━◄

PRITT, THE MAN FROM YORKSHIRE

*N*orth Country Flies, published by T. E. Pritt in 1885, carried on the traditions of both Ronalds and Stewart. The title suggests a regional identity, and Pritt served as its spokesman. North-country anglers found plenty of sport in Lancastershire rivers such as the Ribble and Wyre draining into the Irish Sea in the west. (Also salmon and sea trout fisheries with coarse fish in the lower reaches.) The Coln, a chalk stream and a headwater of the Thames, arises in the Pennines in West Yorkshire, while others in North Yorkshire enter the North Sea. Most notably Pritt tied and fished soft-hackle patterns intentionally to suggest caddis pupae, thus more clearly fulfilling the definition of "upstream nymphing," although that term was never used. Directly or otherwise, Pritt managed, through a colleague, to greatly perplex, confound, and annoy Halford by raising some grave doubts about the dry fly as the only effective technique suited for his chalk streams.

I might have overlooked Pritt were it not for the 1995 facsimile publication of *North Country Flies*. The original was so well known and popular that when it became a rare collector's prize, a company (Smith Settle) in West Yorkshire reproduced the work replete with color illustrations. It's very welcome, too, when a reprinted classic includes a brief biography, in this instance written by historian Leslie Magee. Pritt was born in Lancashire in 1848. His father and an uncle were successful artists, a talent Pritt also demonstrated in his self-illustrated *The Book of the Grayling*. Following school he pursued a career in banking with great success; he became Bank Inspector at Leeds in West Yorkshire and eventually opened his own bank. And in West Yorkshire he found the River Wharfe together with clusters of convivial anglers in various "pub clubs." They were not all fly fishermen by any means, and Pritt was far from a purist. He and three close friends learned from one another, in Pritt's case how to take grayling on a drifted worm—"the acquirement of this most successful and fascinating branch of the Art of Angling." The friendly banker took on an additional responsibility as fishing editor of the *Yorkshire Weekly Post*, earning a wide readership for his eclectic style with elements of fish catching, poetry, and humor. Not yet fifty, Pritt died in 1895 after a brief illness. Today the River Wharfe is one of the best-known freestone spate rivers in the north of England. It and a good dozen other regional rivers sustain large stocks of brown trout, some sizable, and they can be "had" with the dry fly, upstream nymph, or "spiders," now classified as "soft hackles." Pritt's patterns, such as the Orange Partridge, Waterhen Bloa, Dark Snipe, and Purple still hold sway and, through Sylvester Nemes, have a following in America (*The Soft-Hackled Fly*, 1975). Although they didn't fish their soft-hackles in the same way, Nemes still called Pritt his hero!

Pritt would have no doubt been happy to imitate the mayfly, "which sits upon the water like some stately winged queen from fairy land." It was just that the swift rivers bound for the North Sea were not conducive to many hatches of significance. Thicker skinned than Stewart, he wrote, "In the clear, smooth, gliding waters of the chalk streams of Hampshire and a few other counties, the case is different. There, fly-fishing as an art is perhaps at the greatest perfection." Pritt noted that his rivers "have their rise on wild, cold moorlands, high up bare hillsides, many of which approach 2,000 feet." The snow-fed waters remained cold well into the short growing season, and the streams tended to be the shoal, swift and only occasionally graced with pools. As a consequence the trout ran smaller than the hefty chalk-stream specimens, although apparently still weighing around a pound and certainly better than "herring size" as was claimed by the Hampshire crew. Pritt confirmed that spring-fed streams were more challenging and that the only local chalk stream, the Driffield, was exceptionally difficult.

CHAPTER EIGHT

Ronalds's Influence

True to its title, *North Country Flies* was primarily devoted to a listing and description of important Yorkshire patterns. Whereas soft-hackles were sometimes called "spiders," I've no evidence that Stewart's unique tying style was copied, nor does the book take up methods. (It appears that the Stewart spiders were more heavily hackled.) Pritt's preference for hackle flies was simply the belief that that sharp-eyed trout are more likely to recognize counterfeit when presented with flies tied with bulkier materials such as wings, whereas the more delicate hackle suggested a pupa struggling to attain the surface of the water. Thin bodies were tied with silk, a film of dubbing overlying. Following the style of Rolands, the common name was followed by a recipe for the ingredients and comments regarding the insect to which it was "matched" together with suggestions about time of season and weather. Working through the sixty-two flies is tedious in that Pritt constantly refers to Ronalds's similar or identical patterns with different names. And the soft hackles came from an array of unfamiliar birds. Take the Water Hen Bloa. It turns out that *bloa* is merely a Yorkshire name for a dark blue-grey feather, not an exotic fowl left over from the Stone Age. Pritt indicates that the Waterhen is effective when caddis are "struggling to attain the water surface" in cold, windy weather. Mayfly imitations were not entirely ignored, with five dressings for the March Brown. Another, the Little Dark Watchet, was as popular as Ronalds's Iron Blue Duneven in the south. These had wings, but thin as daggers. It should be noted that Halford converted patterns such as the Grannom, Yellow Dun, and Iron Blue Dun into dry flies. They were among his favorites for many years.

Caddis

As in the case of Stewart's Spiders, Schwiebert was convinced that, as caddis populations go, so do soft-hackle patterns. That is, he saw them as structurally efficient imitations of caddis pupae (*Trout*, 1978). Pritt wrote, "Again, there can be no doubt that a large number of the winged transformations of the pupae never reach the surface of the water: either they are imperfectly hatched and are seized as they are borne down the stream, or they are pounced upon by eager trout in their ascent from the bottom of the river to the air. In both cases the hackled fly is quite sufficiently deceptive." In *Nymphs* (1973) Schwiebert discusses the Grannom, a principal caddis of the early season: "No well-stocked wading vest can be without this imitation, which I have adapted from the North-Country flies developed by such British writers as Stewart, Pritt and Scrope a century ago."

Pritt considered the size and color of a fly as most important, but still relied upon the allure of "imperfect shape" mimicking a "half-hatched" insect as contributed by the soft hackle. Pritt normally fished upstream to take advantage of "natural drift." An upstream breeze was greatly appreciated, although in high, discolored water he'd fish downstream, maintaining some slack by wading ahead of the fly. Whether or not the following description is a "first," it has significance: "On very cold, dark days if the hackled fly be allowed to sink in the stream as far down as possible, and then be quietly worked up in series of short and gentle jerks, it will often account for fish which can be gotten in no other way." To paraphrase, Pritt nudged the deeply sunken fly upward in the water column, aping the tentative streambed-to-surface movements of a pre-emergent caddis or mayfly.

Mentioning Pritt in context, Schwiebert noted that anglers such as Edward Hewitt and James Leisenring later adapted the "subtle lessons of British writers to American wet-fly problems." These early lessons pertained to both pattern construction *and* presentation (*Trout*, 1978). It's uncertain whether imitations of caddis larvae were successful. Pritt wrote: "It is worth remembering that trout feed largely off the bottom of the river upon the pupae and larvae which are constantly on the move there. When they are feeding, there is no food on the surface and their eyes are intent upon the bottom." Trout rarely took any artificial fly at such times although caddis-baits sometimes worked.

His Contributions

Pritt's place in fly-fishing history is influenced by the marked similarities in style with Ronalds's *The Fly-Fisher's Entomology*. That Stewart's influence was also strong is unquestioned. Parts of several chapters could have come from *The Practical Angler*, such as those devoted to the "stonefly creeper," the minnow, and upstream worm. He added a bit of detail, such as dropping the worm several feet behind the trout in expectation that the fish would turn to take the worm. (Both Hewitt and Marinaro employed the same tactic with flies.) Nonetheless, through his book Pritt was a major factor in establishing the soft-hackle pattern style and upstream presentation in the northern counties and Midlands at a time when the dry-fly renaissance was in full bloom in the south. Did Pritt have an influence upon Skues, who was just becoming interested in the upstream nymph at the time of Pritt's death? They imitated different insects in different ways in different types of streams. Still, *North Country Flies* certainly advertised the upstream wet-fly method, encouraging exploration. This brings us back to issues with Halford. In brief, a skillful Yorkshire colleague, a Mr. Reffitt, fished with Halford on the Test and did remarkably well, much to the amusement of the great man's critics. This went beyond wet fly v. dry in that Halford absolutely denounced the practice of "fishing the water" or if you prefer, "chuck and chance it," whereas anglers from the north felt no compunction about prospecting when necessary. Skues knew all about the incident. (A full account is given in the chapter concerning Halford.) I give Pritt great credit for documenting the importance of the caddis pupa for fly fisherman. At about the same time, George Pulman, subject of the next chapter, did the same for the adult caddis. Regardless, as history unfolded, first in England and then America, the mayfly increasingly occupied center stage. The caddis remained very much a Cinderella well into the twentieth century.

Pulman Fishes the Surface

*R*ather than becoming embroiled in the captious process of identifying the father of the dry fly, historians have settled on a reasonable and generally acceptable choice. George Pulman's credentials for the honor were layed out in his *Vade-Mecum of Fly-Fishing for Trout* in 1851. Pulman was a respected tackle maker in Axminster, a small Devon market town on the River Ax, several hundred miles west of Hampshire. Arising from limestone caves well to the north, the Ax is a classic chalk stream, winding through farm and pastureland on its way to the Chanel. John Waller Hills said of Pulman, "He produced the theory and practice of the dry fly full grown from the brain of the parent" (*A History of Fly Fishing for Trout*). John McDonald agreed: "George Pulman pulled the dry fly out of his hat complete, though for all anyone knew it might as well have been a rabbit" (*Quill Gordon*). I preferred Arnold Gingrich's view: "There had been frequent mentions of the taking of trout with flies 'on top of the water' going back over nearly two centuries, but the first reference to dry-fly fishing in the sense that we know it today came in Pulman's book" (*The Fishing in Print*). Herd suggests that it may have been Pulman who invented the first adjustable drag mechanism for a reel (*The Fly*).

THE VADE

I looked it up. A vade mecum is a handbook. My well-read hardbound expanded third edition (1851) measures just over six by four inches. The book is an excellent primer for the beginning angler but is hardly meant to be carried on the water for it is neither brief nor superficial and deserves thoughtful reading. Nor does Pulman pay a great deal of attention to the dry fly! He does address tackle, the water forms preferred by trout, how to approach them safely, the rudiments of casting, setting the hook, playing fish, and even a bit on courtesy and conservation. There are chapters on tying and pattern selection. Sprinkled in with the "how to," Pulman pleasantly extolls the beauty of nature and the virtues of the simple, rural life after the manner of Walton and there are snippets of Ovid and other poets. So what about the dry fly? It takes some looking. After six chapters and 131 pages (out of a total of 186) you come to a description of a mayfly hatch, the trout taking duns from the surface.

> So, as it is not in the nature of things that this soaked, artificial fly can float upon the surface as the natural ones do, it follows the alternative and sinks *below* the rising fish, the notice of which it entirely escapes, because they happen just then to be looking *upwards* for the materials of their meal. Let a dry fly be substituted for the wet one, the line switched a few times through the air to throw off its superabundant moisture, a judicious cast made just above the rising fish, and the fly allowed to float towards and over them, and the chances are ten to one that it will be seized as readily as the living insect. This dry fly, we must remark, should be an imitation of the natural fly on which the fish are feeding because if widely different, the fish, instead of being allured, would most likely be surprised and startled at the novelty presented, and would suspend feeding until the appearance of their favourite and familiar prey. . . . it is *not* recommended that a dry fly be substituted for a wet one over every rising fish. It is really only under particular circumstances, and in favorable situations that the motions of the natural insect can be so imitated as to prove successful, unless the fish are ravenous and seize everything presented to them—a state of things not often experienced.

Pulman first got our pulse to racing with that ten to one comment, but then cautioned against getting too excited, not exactly a hard sell. But there are more credentials scattered here and there. A lot has been made of Pulman's advice regarding false casting as proof of his intention to dry his fly. However, according to Herd, Pulman was actually describing a false cast for the purpose for clearing water from a waterlogged line rather than for drying the fly. Herd indicates that the first description of the false cast to enhance the fly's buoyancy didn't appear until 1853 (*The Fly*). Pulman saw no way of tying a floating fly and so depended on fresh, hence temporarily dry, flies.

Why Did It Take So Long?

According to all accounts, fly fishing in England was very productive well into the second half of the nineteenth century. A brace of wet flies fished downstream pretty consistently filled out large bags, and trout weighing five pounds and more were not rare. So practically, there was little motivation for innovation. The slow development of tackle propitious to the presentation of the dry fly, especially powerful, light rods, was an impediment. According to some, the definition of dry fly fishing required tackle capable of delivering the fly effectively regardless of the elements. Pulman fits here, too, having spent an entire chapter on the desirable v. undesirable qualities of the fly rod. When expressing disgust for the supple fourteen-foot rods typical of the day and "the dandy summer-day fly-whipper," he comments, "rod makers are seldom rod users." He preferred shorter, lighter, and stiffer rods sufficient to throw a long line or to cast into the face of the wind, noting, "The tapered, oiled silk line needs to be heavy enough to work such a rod." These remarks pretty well defined the rods that dry fly fishers of the future developed. Pulman continues, "There should be the finest gut at the tip and the lightest hook so that the fly will alight upon the water quiveringly and insect-like. . . . slowly and floatingly, it alights like a gossamer, and as straight as an arrow—I throw, indeed, as if the surface of the water were a yard higher than it really is." The Pulman's credentials are mounting up!

Entomology and the Dry Fly

Following hard on Ronalds's heels, Pulman was a well-informed streamside entomologist. The *Vade* contains solid, practical information about the caddis, mayfly, and stonefly. The caddis actually leads off.

Pulman states that there are "upwards of two-hundred caddis species in Britain," of which the "Grannam" is particularly important. Herein the distinction between cased and free-ranging caddis is discussed. Pulman speaks of the "winged feast" as a caddis dry fly opportunity very early in the day in May and June and in the "shades of evening" later in the summer. His only error, unexpected for such an observant angler, comes when we're told that caddis eggs are laid upon vegetation overhanging the water. Pulman regarded attempts at true imitation as quackery, although he argued for impressionistic patterns conforming to the size, coloration, and form of the natural. He was also a minimalist, stating that the Red Palmer and Blue Dun, both recommended by Ronalds, will often suffice throughout the season. Aside from a preference for smaller lightweight hooks, there is nothing in his fly-tying instructions to suggest modifying materials or methods to facilitate flotation. The text requires some reading between the lines because many of Pulman's comments have a dry-fly ring and yet he doesn't make a clear connection. Reference is often made to traditional downstream fishing with a dropper fly, and various instructions are equally applicable to either the wet or dry fly, such as the recommendation in favor of fishing a single fly on weedy streams in the summer, implying that a brace was still standard otherwise.

THE INSTRUCTOR

Pulman's pleasant sense of humor is displayed when discussing the trouts' sense of hearing: "It was a bountiful day, the fish coming thick and fast, when a band of roving musicians joined him at the edge of the bank, a crowd of followers in tow. All parties, except Pulman, were quite inebriated. . . . half-stunning us with the squeak of clarinet, the clang of trumpets and the thunder of a rolling drum, amid the uproar of a fight among the bystanders. We endured the din with proper equanimity, landing fish after fish at the very feet of our disturbers. . . ." That's even a better experiment than Ronalds's! Pertaining to the caution required in stalking a fish, we meet the aggressive angler who stalks "like a man escaping from a burning building." Whether trout feel pain has recently been debated. Pulman thought they must be pretty stoic, otherwise so many fish with one or several flies in their jaws wouldn't continue to feed. And we can all relate to the following piece of advice: "Crack! Ah! the knell of a departed fly—produced by your whip-like management of the rod. That forward movement was made too quickly—you did not allow sufficient time for the line to unfold itself behind you, and the loss of your front fly is the consequence." Lest the reader become too busy with false casting, he is warned that the flash and movement of the rod may put the fish down. Pulman goes into rapture over the reliability of the Double Surgeon's Knot—"a bond only second to that of matrimony." I liked that. I wondered whether he would have comments regarding fishing pressure or pollution as had Stewart? "The valley is enclosed and sheltered by ranges of high hills which the hand of Cultivation (the arch-enemy of landscape beauty) has not yet quite denuded of their wood, although its approach is everywhere manifest." It's not clear whether Pulman was concerned only with esthetics or if he was referring to the deadly effects of deforestation. He says nothing about pollution or fishing pressure but shows disgust at the shortsighted and unsportsmanlike killing of smaller fish with parr markings. His reason was highly practical, for many local anglers confused juvenile salmon with young trout, killing fish that presumably would have returned to spawn. Pulman's concept of sport was nonetheless much like Stewart's. He refers approvingly to three adept anglers as "slaughterers of the first degree." The Ax today boasts two annual salmon runs and a sea trout run as well as qualifying as a classic chalk-stream trout fishery. The area remains a popular holiday destination.

PULMAN'S CREDENTIALS RE-EXAMINED

It's my understanding that the earliest edition of the *Vade-Mecum* (1841) failed to mention the dry fly. We can surmise then that Pulman's experiences with the method took place during the next decade. Francis Francis, mentioned in more detail shortly, wrote a piece in *The Field* advocating the dry fly in 1857, and since he must surely have read the *Vade-Mecum*, Pulman's influence is likely. Winchester, which had always attracted crowds of anglers, was to become the hub of the dry-fly excitement, but Pulman developed his ideas on his own in a more remote and less populated area, which suggests his originality. In conclusion, the *Vade-Mecum* is a general guide and reference well seasoned with philosophical considerations. The dry fly comes across pretty much as a side note, an occasional method for when conditions were just right. Curiously, in Skues's first book, *Minor Tactics of the Chalk Stream*, published some sixty years later, the nymph is the minor tactic.

Francis Francis Played His Part

Pulman's role in drawing attention to the dry fly was likely less than that of his irrepressible contemporary, Francis Francis. F. F., as he was known to his many friends, assumed the editorship of *The Field* in 1857. According to Tony Hayter, "This journal had a paramount influence in the sporting world of the nineteenth century—a research project into the history of almost any sport in Britain would have to begin with *The Field*." It played an important role in pulling together numerous angling practices and attitudes and moving them towards becoming developed schools or doctrines. For nearly thirty years the writing was mainly done by Francis (F. M. *Halford and the Dry Fly Revolution*).

F. F. had fished widely in England in many ways and for fishes of many sorts. He wrote easily and pleasantly. More importantly, Francis could see, understand, and relate to contrasting viewpoints. Unless poaching or poor stream management was in question, he was more likely to defend than to take sides. Without deprecating the downstream method, Francis wrote influentially of the dry fly in *The Field* and in *A Book on Angling* (1867). Perhaps his main point was that during the calm and heat of sunny summer days when attempts with the sunken fly were futile, the dry fly offered good prospects for success, in effect nearly doubling the length of the season. F. F. became a close associate of Halford and Skues, playing the part of a latitudinarian in the gathering debate while maintaining friendships all around. ◄━━

Marryat, the Power Behind the Throne

*G*eorge Selwyn Marryat had no equal when it came to removing sophisticated trout from the glassy southern chalk streams. Halford simply called Marryat *the best*, without qualification or equivocation. He remains a legend for fly fishermen with an appreciation of angling history. It's rather remarkable that fishing was Marryat's third "career." After graduating from the Winchester School, Marryat entered the military in 1858, following in the footsteps of his father, Colonel Marryat. There is a fly-fishing connection in that in the Dragoon Guard's extensive practice with the saber was required. As Hayter remarked, "His later prowess with the heavy fly rods of the time, and even his ambidextrous casting, must have owed something to this earlier conditioning" (*F. M. Halford and the Dry Fly Revolution*). Captain Marryat, now a civilian, briefly explored the cattle business in Australia in 1865 before returning to Winchester in 1870. Still in its infancy, the dry fly was attracting a great deal of interest, and the ancient city and the Itchen River formed the hub of the excitement. Fascinated, Marryat studied fly tying under a local expert and quickly became an expert himself. Other skills developed so rapidly that within a few years he stood alone among regional anglers. He was ambidextrous, whether casting long or short, or even under one leg. Interested in all facets, Marryat played a major role with Henry Hall in the development of the eyed hook. Attaching the leader to the eyed hook made for less bulk and weight than lashing the tip to the shank—essential to the dry fly. Further, the union was more flexible and facilitated changing flies. Marryat attributes to "a Good Fly-hook, the temper of an angel and penetration of a prophet, fine enough to be invisible and strong enough to kill a bull in a ten acre field" (Hayter, *F. M. Halford and the Dry Fly Revolution*). He also participated in efforts to improve woven, tapered lines.

Well educated and highly intelligent, Marryat was gifted with a sharp wit and a remarkable range of interests. An entertaining companion, he could discourse upon the sciences, the humanities, and even matters metaphysical. Despite his fame, Marryat remained easily approachable. He could often be found chatting amiably with fellow anglers along the Itchen and neighboring rivers, and he gladly offered advice to beginners. As a schoolboy at Winchester, Sir Edward Grey met Marryat along the Itchen and later called him "the greatest angler I have ever met." "One could not say which was the more instructive, to watch his fishing or listen to his talk; no one had more information to give, no one was more generous in giving it; his knowledge seemed the result not only of observation and experience, but of some peculiar insight into the ways of the trout. In the management of rod and tackle he displayed not only skill but genius" (*Fly Fishing*, 1899). When Francis Francis incorrectly blamed his fly for a lack of success, Marryat replied, "It's not the fly, it's the driver." Halford further explains this oft-quoted incident in *Modern Development of the Dry Fly* (1910). Every writer has pondered the extent to which Marryat contributed the substance to Halford's first two books. A historian quoted by Kenneth Robson suggests that "without Marryat's genius Halford would have been very hard pressed to have produced the work he did" (*The Essential G. E. M. Skues*, 1998). Marryat's position and fame were already well established when Halford, the new boy of the block, began to fish the Itchen. The two became fast friends and virtual partners in defining the dry fly in practice and ethic, as enjoyed by chalk stream anglers. This involved a great deal of "research and development" in the study of insects and their imitations. For much of this time Marryat and his wife were living in Salisbury, due west of Winchester, while Halford, a Londoner, spent a great deal of time in Winchester. The men carefully identified a list of dry-fly-related problems, puzzles, and ambiguities that called for resolution. These they attacked together or separately, but always in partnership.

Halford had always intended to publish "for the benefit of the angling fraternity, all details we had worked out. It was always my idea that Marryat would collaborate." Though neither self-effacing nor reclusive, Marryat was curiously reluctant to share his wisdom in printed form and declined. Halford nonetheless profusely credited his friend for all he had contributed, as we can see in this excerpt from the preface to Halford's first book, *Floating Flies and How to Dress Them* (1867): "To this friend, George Selwyn Marryat, I desire to express the deepest gratitude for the unwearying patience and perfect unselfishness with which he gradually inducted me into every detail known to him, and gave me the benefit of his invaluable experience, concealing nothing which would tend to perfect me in the art of imitating the various winged inhabitants of the streams."

Marryat left us no books, published only a few brief articles, and was uneasy when asked to speak publicly or even to attend large sporting banquets, where he feared being lionized. For this reason Marryat's angling history, covering some twenty-four years, is largely immersed in Halford's. Marryat's untimely death in 1896 shortened their collaboration by a dozen years. Unfortunately, when it was suggested that Halford write a biography of his friend, he disclosed that he had failed to preserve Marryat's extensive correspondence (*F. M. Halford and the Dry Fly Revolution*). Perhaps it didn't matter, for Marryat's fame had already spread far and wide. He was virtually deified, fly fishing's avatar. The American angler Theodore Gordon, an icon in his own right, never visited England, but had this to say: "What a wonderful man Mr. Marryat must have been, and what a privilege it is to meet such occasionally in our rambles through the world. Kind, genial and generous, with the innate nobility and strength of character which we presume should be attributes of all men in human form divine" (*The Complete fly Fisherman: The Notes and Letters of Theodore Gordon*, 1970).

Halford, the Messiah

*E*very sport has had its heroic figures, Olympian perform-
ers of extraordinary strength, speed, agility, and courage.
Frederick Halford's legend is of quite a different stripe.
Marryat was by far the better technician, and when fishing with other friends, Halford didn't necessarily come out top rod.
Think of him instead as the leader of a religious faith centered at
Winchester with an apostolate scattered throughout Hampshire
and neighboring counties. Members of his congregation sought to
achieve redemption and grace through their devotion to the won-
drous chalk-stream trout. And it was Halford who defined how mass
and confession should be celebrated. His were not rules or laws pun-
ishable by fine or imprisonment but rather pertained to propriety and
ethics, fly-fishing's moral high ground. The extent and duration of
Halford's influence was reflected by Mr. Gordon, my elderly golfing
friend from the 1950s, a loyal American "Halfordian." The label has
an aristocratic ring.

How did a single individual manage to rise to ex cathedra eminence and recruit such a far-reaching and lasting following? While
I've managed to collect all but one of Halford's books, my insights
into the man would have been far less complete were it not for Tony
Hayter's *F. M. Halford and the Dry-Fly Revolution*, which collects correspondence that often allows one to read between the lines.

Fishing the Serpentine

Born in 1844, Halford was a true Victorian in that his life closely coincided with the reign of the Queen. Halford's extended family pioneered the ready-made clothing industry in London, gaining considerable prosperity. His formative years predicted his later character rather accurately. Even as a child he displayed such a fixation upon fishing that family and friends poked fun. Growing up, he came to expect it. It mattered not where and for what he fished. Perch in neighborhood ponds and streams gave way to expeditions to the Serpentine in Hyde Park and the Thames where barbel, bream, and roach were the prizes. Following completion of his education and still in his late teens, Halford was employed in the family's company offices, devoting his vacations to fishing. In his twenties various ocean species became a focus and then, baiting for monstrous sea-run browns in the Thames. Always intense, Halford began to display a precocious interest in innovation that led to the development of better terminal tackle for probing the river bed, where he caught bream of such dimension that they were preserved in a glass case in his father's home. He was pampered as a child, and his good fortune continued as evidenced by a private guide and boat on the great river where a trout weighing just under seven pounds fell to his bleak. (Bleaks are small, pelagic bait fish related to carp.) Halford's description of the Thames was one of serenity and beauty as it ran through the verdant, largely unspoiled countryside. River traffic was light and only an occasional river barge passed though the primitive wooden locks (*An Angler's Autobiography*, 1903).

Attempting the Dry Fly

Privilege looked again upon Halford when in 1867 an old friend of his father's invited him to fish private water on the lovely Wandle, a chalk stream not far from London. Finding that fishing wet flies downstream was futile, he met other anglers experimenting with the new dry-fly method, "opening out his mind to an entirely fresh vista of a novel form of angling." On a visit to a leading London tackle shop he procured an eleven-foot hickory rod, one of the new waxed silk lines, and a selection of flies best suited for the swift streams to the north. Eventually though, the Hare's Ear, for a time his favorite pattern, began to take some fish from the Wandle. Here Halford found a "honey hole" where trout often rose just below a bridge. As it happened, however, the limbs of a tree shut off all casting lanes. After arranging for the tree branches to be removed by the gardener, he killed a nice mess of trout to be shipped triumphantly to London. It should be mentioned that segments of many streams were set aside for the "Commons." Here the fee for the rent of a rod, a ticket to fish, was quite reasonable. However, Halford observed, showing his patrician side "the weeds are not cut, the pike are not killed down and it is no one's business to see that those fishing either have or rent rights" (*An Angler's Autobiography*).

The Houghton Club on the Test

By the 1870s, decent fishing within easy reach of London was becoming harder to find and more expensive. In this seller's market, Halford found it necessary to preview various beats put up for rent lest they be short of fish and polluted with sewage. And so in 1877 when an opening appeared for membership in the ancient Houghton Club at Stockbridge on the Test, he didn't hesitate. (Stockbridge is a few miles northwest of Winchester.) Formed in 1822, this was an assemblage of twenty members "all drawn together by the irresistible magnetism of the grand river Test." The dues were twenty pounds per annum. Halford soon discovered that he'd previously been fishing in the minor leagues. Until I made my earliest attempts on this historic Hampshire stream, I was really under the impression that I knew something of dry-fly fishing and fancied myself rather a good hand at it. I was quickly disillusioned. The well-fed chalk-stream trout had become angler-wary and closely examined imitations of an array of mayfly species drifting leisurely on the clear, slick currents.

Although cautious and unassuming, Halford was determined to improve his knowledge and skills. Thus meeting Marryat and other leaders of the dry-fly movement was imperative. Victorian relationships were formal and reserved, however, and familiarity only developed gradually, over time. Although he finally met Marryat in 1879 at Hammond's fly shop in Winchester (a Mecca of sorts), it was some time before they fished together. Halford's dedication and effort coupled with his pleasing personality resulted in eventual acceptance into the inner circle of chalk-stream regulars.

This was a period of change for the Houghton membership. The older members included a good many downstream, multiple-fly anglers. Now they were gradually being replaced by younger members of the "up to date school," as Halford put it. Tony Hayter points out that advances in the physical and biological sciences were being applied to develop new insights into the solution of traditional angling problems. It was a revolution or renaissance of a sort, as the title to Hayter's biography suggests. Halford and many of his friends were major players. Francis Francis, whom we've met earlier, was large in person, in appetite, and in companionship. He presided over The Mayfly Mess, an annual celebration during the Green Drake hatch held at the "court" or Royal Hotel in Winchester. The Mess was an exercise in food, drink, and fellowship enjoyed by an assemblage of anglers of similar persuasion. As mentioned, Henry Hall, engineer and school teacher, worked closely with Marryat in improving light wire up-eyed hooks and was instrumental in gaining their general acceptance. Hall wrote several books on the dry fly before Halford's as well as producing an algebra text that served as a standard well into the next century. H. T. Sheringham later took over the editorship of *The Field* from Francis Francis and was accordingly an interested observer in the upcoming Halford v. Skues unpleasantries. R. B. Marston founded the prestigious Flyfisher's Club in 1884 and was editor of the *Fishing Gazette*, another major organ. He published Halford's first book. Dr. Tom Sanctuary, another member of the fly-designing and hatch-matching consortium, found that oil from a duck's preening gland improved floatation. Major William Turle is best remembered for the "Turle Knot" used to attach fly to gut. It was still in popular competition with the Clinch Knot when I was a boy. Major Anthony Carlisle provided useful information by recording correlations among the weather, water conditions, insect activities, and the fishing. He later became the club's historian for that eventful period. Martin E. Moseley, Halford's nephew, played a particularly important role. Robson indicates that he was employed by the Natural History Museum and was regarded as the leading entomologist of his day. Mosely authored *Dry Fly Fisherman's Entomology*, serving as final arbiter for Halford and Marryat when uncertainties regarding classification arose (*The Essential G. E. M. Skues*, 1998).

Reminiscent of Cotton's hut on the River Dove, Hayter mentions an early opportunity for Halford when he spent a day sheltered from the elements in the Sheepbridge Hut on the Test in the company of Marryat, Hall, Francis, and Carlisle. He had been honored to be part of a "Hampshire Summit" (*F. M. Halford and the Dry Fly Revolution*). With the exception of Marryat this list of associates sounds like a consortium of writers and editors. These men were all skilled anglers as well, contributing and testing new ideas. Besides, a revolution needs to be advertised. Through his books and many, many contributions to popular journals, Halford was on the front cover of the prospectus. Ever thorough and meticulous, he was well suited for the business world. I wonder, though, how he resolved the stark contrasts between the tidy, dusty details of business ledgers and the call of the fresh-air applied biology that's so much a part of fly fishing. For him the dry fly had become another "business," a serious one and, as we shall see, eventually more important than the other in London.

Floating Flies and How to Dress Them

Halford and Marryat considered the commercial mayfly imitations ill-conceived and poorly constructed. Thus their first attempts were to develop tying methods that would facilitate floatation while greatly enhancing realism. Published in 1886, Halford's first book, *Floating Flies and How to Dress Them*, is short in pages, but characteristically long on detail, clearly set forth. The considerable credit given to Marryat no doubt helped to attract readers. A great deal of attention is paid to various dyes for obtaining precise tints of color. Some were the same as Ronalds used, and aniline dyes had also become available. The complexity and sophistication of certain patterns is impressive, paired wings cocked just so and long, detached, upward curving "mayfly" tails. Well-executed drawings suggest that the flies were tied without benefit of a bobbin, the silk kept under steady tension with one hand. Half-hitches are dismissed as unnecessary and bulky. Halford paraffined his flies before putting them in boxes. A catalogue of some eighty-two patterns follows. As with Ronalds, concise comments as to when and where they might be most effective are appended. Only twelve of Ronalds's patterns appear as a reflection of Halford's preoccupation with mayfly imitations. At the very end, almost as an afterthought, the reader finds just a few words regarding presentation. When Halford mentions that another work is in progress, I correctly assumed that the second volume was intended to flesh out his dry-fly method.

Dry-Fly Fishing: Theory and Practice

This much weightier book appeared three years later, dedicated to George Selwyn Marryat "by his grateful pupil." The reader is delighted to find a chapter on chalk-stream structure and where feeding trout are likely to be found. Others speak to the feeding habits of trout and grayling, how to select the proper fly, how, when and where to cast, set the hook, and play a fish. Halford also gives instruction and advice about topics that would never have crossed my mind. Have you ever wanted to learn how to properly autopsy a trout (diagram included) and then sort out the gastric contents into surface insect remnants and. bottom creature bits? Did you know that porpoise hide makes the best laces for wading shoes? I found out too how to get rid of predatory pike and also poachers. There is remarkably little repetition or overlap with the first book. An exception is the insistence that the first cast be as perfect as possible, the fly alighting upright and its wings cocked. (This is almost a mantra.) The chapters on casting are typical of Halford. Take a topic, dissect it into its component parts, define and name each piece, place the components in some order, and discuss them exhaustively. Thus we meet five techniques for propelling line and fly, each designed to best address a particular situation.

Halford discusses strategies too. He avoids racing other anglers to the best water, preferring to hang back until the stream has had a chance to settle, and bypasses the most popular pools and runs in favor of more technically challenging water where he always seeks larger trout and grayling.

A team of three photographers and assistants spent a week with Halford capturing the talented Marryat in informative poses, the line frozen in a critical position. Clad in knickers and sporting his signature tam, Marryat cuts a trim figure. In some casting stills, his left hand appears to be idle, hanging by his side. (We're told that trout are to be "struck" from the reel.) Posed portrait photos of Marryat elsewhere reveal a puckish countenance and rather delicate features more suggestive of a poet or artist than a former military officer and cattle raiser.

Then the Catechism!

Halford offers just three dicta in this book: First, the dry-fly man must not bring his rod to life until he has spotted a fish taking insects from the surface. Second, that surface disturbance or "spot" is then cast to "to the exclusion of any chance work in other parts of the stream." Third, if a trout or grayling refuses three or four well-presented cocked flies, they should be rested, perhaps to be re-tried later in the day. Halford insists, over and again, that the more a fish is cast to, the more likely it is to become shy and therefore difficult or impossible to catch. Halford has a tendency to see human emotions such as shyness in trout. At first this struck me as odd. But in that era horses still ruled the streets and highways and were known to "shy," so it makes sense. Other anthropomorphic attributions include the capacity for curiosity, concern, confidence or lack thereof, greed, irritability, becoming scared, and, when unusually easy to catch, silliness. Once in "too long an awhile" we enjoy a day when the silly trout take our flies with reckless abandon for hours on end. Unable to explain the phenomenon, Halford calls these "happening days," I suppose miraculous. (Stewart said they were *like angels' visits, short and far between* [*The Practical Angler*]). There were also certain trout behaviors that Halford greatly disliked. He didn't care for "bulgers" (his term) that raced and darted about, even bulging the surface. Noisy, splashy risers were an annoyance and lastly, "smutting" fish, taking midges. Why? In each case the dry fly was ignored, thereby depriving the angler of his sport. These were delinquent, roughneck fish, representatives of a lower caste.

Begrudging the Wet Fly and Some Backing Off

Halford's three restrictions are pretty profound, for on those days when the fish are down, the dry-fly angler can only walk the banks, rod at the idle, enjoying the beauties of nature while thinking deep thoughts. Dry fly fishermen "fish the rise" while their counterparts "fish the water" and only cast to a riser when one is handy. Halford gracefully acknowledges the skills of a first-rate performer with the wet fly, referring to Stewart and other north-country upstream anglers. These pleasantries notwithstanding, he clearly believed that fishing the rise required more skill than the "chuck and chance it" approach and naturally showed a strong bias for dry flies in the case of placid chalk streams on bright summer days. He boasted of many converts who succumbed to the excitement inherent in the dry-fly method and even suggested that the floating fly might win out during a hatch in wet, blustery weather.

Surprisingly, Halford ended with a couple of stark contradictions. First, he bemoaned the increasing popularity of the method he helped develop and foster; "the spread of dry-fly fishing has become something dreadful to contemplate." The fishing was becoming more difficult and expensive. Virtually all of the quality Hampshire waters were already privately owned or leased to clubs or individuals, many of them wealthy Londoners. Halford's concerns went deeper. He believed that certain trout have a proclivity to take surface food, whereas others, by nature, are bottom feeders, and that the former were either being killed off or becoming stressed and fearful. Halford even suspected that there might eventually be a genetic shift in favor of the streambed feeders. The Green drake hatch is given as an example. With news of the first signs of the much-anticipated "May-Fly fortnight," hordes of anglers descended upon area streams as the trout begin to feed ravenously on the active nymphs. Halford felt that a true sportsman should refrain until the adults became prevalent and expressed fear that certain stringent votaries of the dry fly were "riding the hobby to death" by adhering to the very stricture he had offered a few pages before. Contradicting himself again, he goes on to recommend, "I for one am a strong advocate for floating a cocked fly over a likely place, even if no movement of a feeding fish has been seen there." Visible trout and grayling, resting quietly, were also fair game (*Dry Fly Fishing: Theory and Practice*). My take is that these reversals were a reflection of Halford's natural caution, as if he wanted to leave a little wiggle in the equation, the window of practicality open just a crack. In 1889, at any rate, Halford was quite capable of violating his own rules. That would change.

The Wet Fly Man from Yorkshire and Some Snickering

Halford held that traditional downstream wet fly fishing practiced on a chalk stream was unsporting. Flogged by anglers walking the banks downstream, the trout could only become shy. Worse, after being pricked by the hook, they were thought to mope forever on the bottom. Halford advised that regulations forbidding the downstream wet fly were to be applauded, for just as on the golf links, etiquette should be observed! Secondly, he assured his readers that the wet fly would not be effective in the chalk-stream setting, whether fished up or down. (This in spite of the fact that he and Marryat had had some success with detailed imitations of Green Drake larvae. They gave the practice up when too many fish threw the hook.) Then at some time in the early 1890s Halford played host to Mr. Reffitt, well-known angler and presumed colleague of Pritt's, spending several days on the upper Test with him. The following account comes from *An Angler's Autobiography*.

> I was much interested, some years since, watching a first-rate wet-fly man, a Yorkshire fisherman, on a portion of the Upper Test. His flies were olive quills of various shades, iron blues, red quills and such patterns, all of which he used on his native streams, and were dressed with peacock quill bodies, very meager upright wings, and a single turn of hen hackle for legs. He did not in any way practice the "chuck and chance it" plan, but moved slowly up-stream, carefully studying the set of the current and quickly deciding where a feeding fish should be in each run. Sometimes it would be close under the bank, sometimes on the edge of a slack place, and sometimes on the margin of an eddy. Whenever he had made up his mind as the most likely spot, there he would make one, or at the most two light casts, placing his fly with great accuracy and letting it drift down without drag. Now this I take it was the best possible imitation of the work of a dry-fly fisherman, except that he had not spotted the fish and his fly was not floating in the dry-fly sense. His patterns were very similar in size, colour and form, to those of the ordinary chalk stream fisherman. He used very fine drawn gut, and worked hard from morning to evening, never passing over a likely place without putting a fly into it and very seldom losing a hooked fish. [This was in the early part of April, during strong, south-westerly winds, when the hatch of duns was sparse; Halford noted that all conditions were "favourable to the sunk and unfavourable to the floating fly."] He fished six days on a well-stocked reach of the river and killed in the aggregate seven trout weighing 9 lb. Candidly, I was somewhat surprised at the good result, and

have often wondered whether he could repeat the performance. Of course the average weight of his fish, just over a pound and a quarter, was very small for the Test, and two or three of them would have been returned by many dry fly fishermen. Let it be clearly understood however, that this fisherman was most skillful and painstaking, and was a past master in the art of selecting the right spot, and in placing his fly accurately and delicately *there* at the first attempt. Had he merely fished the river up or down, or had he bungled his cast, or moved about rapidly, or, in fact, made any mistakes, I do not believe he would have killed a single trout, so that his bag represents the best possible result, under existing conditions, for a wet-fly fisherman on a stream like the upper Test.

Here Halford revealed an unusual mix of emotions: surprise, admiration of a sort, but finally a suspicion that the event might have been paranormal. It's hard to measure the impact the Yorkshire man's accomplishments may have had at the time. The story did get around, though, and in a sense helped set the stage for the debate soon to follow. (Perhaps feeling a need to respond, tit for tat, Halford describes an experience when he turned the tables, catching trout on dry flies in Yorkshire during a low water period when the local soft-hackles made a poor showing, *An Angler's Autobiography*.)

In earlier times, when the chalk streams were necessarily plied with sunken flies, remarkable catches were recorded with downstream techniques and even "dibbling." While Halford was highly respected, there was still a pretty general feeling, even on the part of Marryat, that his ultra-purism was a bit over the top. Marston and a number of other dry-fly apostles fished wets very effectively on occasion, upstream or down. Hayter mentions that they and other dry-fly men, straying temporarily, kept pretty quite about these peccadillos. Members of the Houghton Club were traditionally broadminded. Their chronicles went so far as to record two trout of prodigious weight, one caught on bread, the other on a hunk of meat! Hayter suspects that Halford would have "regarded such misdemeanors with amusement." Sheringham delivered some gentle jibes and eventually had some good things to say in print in praise of nymph imitations. William Senior also wrote a piece *The Wet Fly on a Chalk Stream* for *The Field* in support of the nymph. Halford was seemingly unaffected. He had made plain his views. The matter required no further discussion. Halford responded to sharper attacks from "his critics" without a great deal of rancor, although there is the sense that reactions such as "myopic," "tunnel-vision" or "Autocrat at the water side" must have stung. Other criticisms were rather cruel, among them a number of cleverly conceived lampoons that made great fun of the purist (F. M. *Halford and the Dry-Fly Revolution*). Certain of his doctrine, Halford was not contentious and had no desire to become mired in debate. To quote Hills, "Halford considered that the dry fly had superceded for all time and in all places all other methods of fly fishing and that those who thought otherwise were either ignorant of incompetent" (*A History of Fly Fishing for Trout*, 1921).

HATCH MATCHING

In Halford's attempt to solve the mysteries of chalk-stream entomology, the master plan was to capture and identify as many mayfly species as possible, examine them minutely with the aid of a microscope, and design imitations reflecting conformation, tint, and shade. It was work that might have led to a Ph.D., and Halford's *Dry-Fly Entomology*, published in 1897, was a thesis of sorts. The culmination, of course, was the on-water testing, which was done in collaboration with Marryat. Duns and spinners, males and females were picked from the water or vegetation, studied, classified, preserved, and permanently mounted in Canada balsam for posterity. Rather than turning to Ronalds's scholarly work, Halford preferred to depend upon his own and Marryat's observations, with advice from Martin Mosely. Although in general Halford seemed to believe that refusals had more to do with "shyness" than selectivity, he updated the 1886 list and shared the "hundred best patterns." A hundred is a lot in that he disparaged continually changing patterns, believing that a cautious stalk and delicate presentation were usually more critical. Hampshire dry-fly men knew all about the confusion caused by masking and mixed hatches, so there must have been times when experimentation was necessary. I assume that Halford most often felt that he could accurately identify the insect of interest, thus selecting a fitting imitation. Some chalk-stream anglers used binoculars, and, as you'd expect, Halford discussed the merits of a range of magnification v. field size.

Despite the extensive work that went into *Dry Fly Entomology*, the book missed its mark with respect to the typical man on the stream. The first part featured black and white line drawings of naturals, heavy on mayflies. The middle section featured color reproductions of the "hundred best" patterns, and the work concluded with a long list of dressings. More to the point, there was just too much information and detail for the angler who'd rather be fishing. Halford's self-confidence concerning the choice of the patterns best suited for a particular mayfly hatch was a bit annoying too: "Fortunately I am able to be more precise in these matters than many of the brethren and to base my advice in these matters on the solid foundation of facts, having carefully kept a diary of each day's fishing for some ten seasons and recorded in every case not only the fly with which each fish was killed, but also some particulars of the natural ones on the water, the state of the weather and other useful details" (*Dry-Fly Fishing: Theory and Practice*). Patronizing? I suppose so. But when confronted with such a huge body of compulsively accumulated data that no one else possessed (except Marryat), we can accept some arrogance.

Marryat's Last Years

From time to time and for any of a variety of reasons, "H&M" relocated their home waters. With the collapse of the time-honored Houghton Club in 1892 the two friends and other former members were left "out of water." The resourceful Marryat identified a possibility on the Ramsbury Estate on the Kennet River, a tributary of the Thames. Halford and two others liked the prospect and rented the fishery for the coming season. The water looked promising even though it had been neglected. The group's efforts to restore the fishery are discussed at length in *The Making of a Fishery* (1895). Pike were a problem and dace too. Literally thousands were extracted during the next few years. The weed beds were cut radically, right down to the streambed. Attention was devoted to various water meadows essentials such as improving bank structure and maintaining the flanking irrigation channels or carriers. A "stew" or hatchery on the estate's grounds was put back in order for the benefit of several imported strains of fingerlings. Halford strongly favored stocking to maintain the quality of sport, and the mixing of the wild trout with "stew grown" or hatchery fish seemed a perfectly reasonable, indeed necessary practice. Hayter indicates that although Marryat lacked the funds to buy a partnership, he was a frequent guest and active participant in all that went on. Tragically, the 1895 season was to be his last. He died of influenza in 1896. His death cast a pall, and the Kennet arrangement ended. "At the age of fifty-six," Hayter wrote, "he still had the vigour of a young man, seemed impervious to fatigue, and was fond of spending long hours in winter weather pursuing snipe." Halford in his eulogy wrote that Marryat had become a dear friend to all even "the keepers, every labourer, even, I think, every one of the numerous poachers in the village" (*F. M. Halford and the Dry Fly Revolution*).

More on Fisheries Development and Management

Halford's approach was hands-on as well as academic. The annual Grannom emergence on the Test around Stockbridge was an event much anticipated by the regulars. But for whatever reasons, there were no Grannom in either the nearby Kennet or Itchen. Meanwhile, the caddis celebration on the Test was beginning to dwindle. On this account Halford and colleagues transported numbers of fertilized eggs and also larvae to the other rivers. A hatch followed the next year, but breeding populations failed to take hold and the attempt was abandoned.

There is no better example of Halford's passion than his efforts to better understand the life cycle of the Mayfly. He "secured the eggs from about 120 impregnated females and with an enthusiastic friend, tried to hatch them in captivity in an aquarium." With the aid of his microscope he counted the eggs—as well as could be calculated, about 6,500 per female. In order to facilitate the experiment, Halford obtained mud and waterweed from the native stream. Although many eggs hatched and the larvae burrowed into the silt, none survived to become adults. The London water was suspect. Halford's drawings of insects were as precise and accurate as you'd expect to see today, for example illustrations of the Green Drake eggs and earlier and later stages of the larvae. The same was true of his line drawings of adult mayflies (*Dry Fly Fishing: Theory and Practice*).

The tiny fly-fishing lobby could do little about the various forms of pollution that accompanied expanding population centers. Halford wrote mournfully about the demise of the pretty and once-productive Wandle, mentioned earlier. Today the area is a London suburb with its own underground station. Unfortunately the process of civilization, which has converted country lanes into long streets of suburban villas and fair flowing streams into foul muddy ditches, is still developing and will go on developing for all time, until some future generation of fly fishermen will find it impossible to pursue their sport anywhere within a moderate distance of great towns or cities.

Halford also recognized the potential threats to public health associated with sewage disposal. While he was fishing in Germany, an outbreak of cholera led to the release of carbolic acid into the river. He thought that an over-reaction since it killed all the fish, but conceded that if "an epidemic of typhoid or similar disease should break out in Dorchester under present sanitary conditions the loss of life would be too terrible to contemplate" (*An Angler's Autobiography*).

CATCH AND KILL

Concerns over the consequences of removing quantities of larger trout from a fishery seem to have surfaced very slowly. Stewart's livelihood depended in part upon bringing in large bags, but it seems curious that by Halford's time, when fly fishing had largely become a gentleman's sport, there had been so little connection between catch and kill and the necessity of introducing heavy stocks of hatchery trout to sustain the fishery. The 1893 season on the Kennet, mentioned above, started with an exceptional Mayfly hatch and spinner ("spent-gnat") fall. The trout welcomed natural and artificial with equal enthusiasm. Halford scored eighty-four trout weighing 134 lb., 4 oz. in just a few days. Halford's partners and many guests had equal success well into the summer. Some decent fish were even returned during particularly bountiful days. The 1894 season was eagerly anticipated, but to the group's great surprise and disappointment, the Mayfly hatch was scanty and the trout rose poorly to the summer hatches of smaller mayflies. Much of the sport was restricted to fishing sedges in the evenings. That set a pattern that continued through 1896. There could have been a number of other factors, but the massive 1893 kill was never mentioned.

It's hard to understand the Victorians' insistence upon recording the weight of each victim down to the ounce. Was it a form of competition akin to our bass tournaments? That comparison breaks down in that, to my knowledge, individual totals were never brought forward to be matched competitively against others.

The killing of trout of ordinary dimensions might be explained by Halford's untested and apparently unchallenged contention that a trout, once caught and returned, would seek solace, feeding on the bottom ever after, and could never again be deceived. As to taking the best fish, writers today often compare themselves to hunters— patient observation in locating a trophy, a careful stalk, the shot or cast, pinpoint accurate, and with just the right bullet or fly. Halford, Skues, and many others were hunters, restricting their casts to situations when a good fish (feeding) was dead in their sights. But did they fail to understand that the finest fish in their bags also constituted the cream of the breeding stock? The following remarks from a piece Halford wrote for the *Field* in 1899, suggests that upper size limits had not yet been thought of.

> Twenty years ago when Test fishermen were not numerous and did not fish hard, it is possible that the so-called natural increase of the trout in the river was sufficient to prevent any marked diminution in the stock. The majority [of sportsmen] are adepts with the dry fly, and spend every moment they can spare from their profession or business on the banks of the stream. The result naturally is that a very large proportion of the feeding fish are killed. I would urge most strenuously on all true sportsmen the manifest advantages of not only abstaining from killing palpably undersize fish, but also of putting back all that just reach the twelve inch limit and many slightly in excess of this size (*F. M. Halford and the Dry Fly Revolution*).

English river keepers adhered to the principle, borrowed from their German counterparts, that it was best for a fishery to "keep the stock moving," that is, to take the older fish to allow room for growth of the smaller trout lest the population deteriorate in numbers and quality. An anecdote from his *Autobiography* underscores Halford's inability to separate the killing of fish from his sport. Through a friend he was invited to spend a day on a certain millstream that held numbers of good trout. The host indicated that he'd like to follow the great angler along as an observer and asked that he spare any of his pet fish, presumably hand-fed. When the first fat brownie was hooked, the miller became upset, insisting that the fish be returned gently. Of course it was. Then the second and third fish, taken further from the house, turned out to be pets as well. Finally Halford tried a remote side stream, only to hook more pets! He put up his rod and waited unhappily for the train back to London. His "sport had been spoiled" for the day. Similarly, he attested to the pain he felt when forced to return large trout and grayling caught out of season. To Halford's credit, he did offer helpful advice about releasing fish. They should not be dropped or thrown and he suggested how to resuscitate trout that "feel sick." Although the "bagging" of most trout of reasonable size continued well into the twentieth century, measures designed to preserve the quality of sport were gradually being imposed. Seasons were defined, minimum size limits imposed, and, for some waters, catch limits. There were other salutary customs and rules such as the release of females during breeding season. You wonder though, during the grayling hecatombs, how so many pounds of fish were disposed of—family, friends, the river keeper, or poor folk? Were trout smoked or dried? That's never satisfactorily explained.

The "New Patterns" and the Ultra-Elitist

Hayter indicates that the years just at the turn of the century marked a lull in Halford's career. Marryat was still sorely missed. There was little new on the horizon, and Halford seemed to lose enthusiasm, writing and fishing less frequently than in the past. Without providing details, Hayter remarks that he had undergone surgery and opines that his health had failed somewhat. Accordingly, Halford began to look for fishing more easily accessible than that on the Test, where a good deal of walking was necessary. Then in 1902 he met Edgar Williamson, a successful businessman, who held prized water on the Itchen in the outskirts of Winchester. Halford was delighted to rent a rod at 100 pounds for the season. Williamson was intensely interested in "exact imitation" and re-energized Halford, actually pushing him even beyond the limits of his usual compulsive precision. They began to work together, collecting, tying and experimenting, something in the manner of the Marryat days. Although Williamson died in 1904, Halford continued to test, tweak, and perfect his patterns, matching each to its living counterpart, and began to fish his new patterns exclusively (*F. M. Halford and the Dry-Fly Revolution*).

The culmination of these efforts is revealed in *Modern Development of the Dry Fly*, published in 1910. After checking the availability and price of the book, I nearly decided to leave a gap in the Halford section of my shelf. A first edition in fine condition can get into the thousands! There are nine color plates depicting each of the thirty-three patterns in the series, between two and five flies per plate. (The originals were tied by Farlow & Co. in London, per Halford's instructions.) In addition, another thirteen plates provide the tyer with color guides to the body, wings, hackle, and tails as suggested by Halford. Each page has four panels in shades from lighter to darker. Some such as "Dead Leaf," "Putty Colour," and "Snuff-Brown" would be hard to relate to without this aid. I don't know how many copies of Modern Development were printed. Andrew Herd comments that the popularity of the Halford series was not long-lived in that the Halford definition of exact imitation had more to do with the angler's view than that of the trout (*The Fly*). It may be that the 1923 edition of which I have a copy was the last.

In Part I of *Modern Development of the Dry Fly* Halford greets and reassures the reader in his familiar style, announcing that during the past seven seasons he fished the new patterns exclusively. "I have no other flies in the boxes carried at riverside. Looking back and considering the question in the most judicial frame of mind, I cannot find that I have on any occasion been placed at a disadvantage by limiting myself to their use. On many days, and under varying climatic and other conditions, I am clearly of the opinion that the natural appearance of the artificial flies has largely contributed to any modicum of success I may have achieved with the shy trout and graying of the pellucid chalk streams." The intention was to equip the dry-fly man with a reliable set of perfect patterns to accommodate almost any conceivable hatch. He need only identify the object of the trout's affection, say a "Pale Watery Dun Female," and tie on a #15! That's right, the patterns were named after "their" insect and numbered! "His new system of nomenclature had an austere clarity about it," as Hayter observed; "every artificial simply had the same name as the natural it represented. Indeed so hard had he labored for accuracy in the New Patterns that for him natural and artificial had blended and become almost the same thing" (*F. M. Halford and the Dry Fly Revolution*).

Part I explains why a particular insect deserves to be imitated and how that should be achieved. (Twenty-four of the thirty-three

winners are mayflies.) There's a nice general section on tying instructions with line drawings as well as the specifics for each pattern. Part II
is more a justification of the New Patterns than a defense although
the first chapter is forthrightly titled "Criticism of the Patterns."
Halford realized that widespread acceptance of his New Patterns
could hardly be expected. "The advantages of being able to carry
thirty-three patterns in all instead of a hundred or more, must appeal
to every dry-fly fisherman. He cannot however be expected to pin
his faith on the new and discard the old standards." Farlow and other
shops had carried the "Halford line" for several years by 1910, and,
with commendable honesty, Halford reported on their acceptance.
There were those who "approved very highly, adopting them forever," another group "utterly condemned them," while a majority
"suspended judgement." He could easily have put a more favorable
spin on the matter but did enlist Marryat's posthumous support:
"Had he been spared to this day there is no doubt in my mind that
he would have warmly approved the new patterns, and very possibly
would himself have abjured the use of any others."

The remainder of part II is taken up by a series of "on-the-water"
testimonials demonstrating the effectiveness of the Spent Gnat, Olive
Duns, Olive spinners, Pale Watery Duns and spinners, Iron-Blue
Duns and spinners, and Blue-winged Olives and Sherry spinners in
sequence. With few exceptions, after setting the stage and identifying
the insect, Halford proves that he has its number, literally, catching
sizable trout and grayling to three pounds. But the previous favorites,
such as his once reliable Hare's Ear, don't get a fair shake. It's like the
old story about the judge who, after hearing one side of an argument,
declines to hear the other in the interest of avoiding any confusion.
In one instance a friend of Halford's hooked three good fish on a Tup's
Indispensible, a fancy fly of the type Halford greatly disliked. When
the fellow began to extol the virtue of the Tup's he was handed a male
Black Gnat (#26). Sure enough, his friend caught two nice trout with
it—but so? These accounts of happenings on the stream are rather
informal and frequently include the activities of a "keeper," charged
with netting fish among other duties. The keeper is hardly mentioned
in Halford's other books or those of Skues and yet the paring of each
angler with a keeper seems to have been as routine as a golfer with a
caddy. There is some banter. "Keeper: 'That rise is at your fly.' Halford
(irritated): 'Yes I know it was.'" On other occasions Halford joins

friends along the bank as a spectator, quite relaxed and enjoying the conversation. He speaks of "yarning," the exchange of tales of uncertain veracity (*Modern Development of the Dry-Fly*).

RECOGNITION FOR THE SEDGE!

It's pleasing to discover that Halford could change his mind, even on settled matters and late in his career. Part II of *Modern Development* concludes with a chapter on caddis.

> Any one of the old school of dry-fly men, if asked his opinion of sedge fishing would probably reply that he rated it as perhaps the easiest and least sporting style of chalk-stream fishing; and very likely would add that, as the trout of the Test and Itchen do not feed on these caddis flies until very late in the evening or after dark, it should not be considered a test of skill on the anglers part to delude them into taking his artificial flies. I confess to having been myself in the past an exponent of this theory and should have been a firm adherent to it to this day but for experiences I had in 1907.

Here he recounts the capture of several good trout taken on a Small, Dark Sedge (#31) during the daytime. Encouraged, Halford enjoyed similar success in the evenings although admonishing that it would not be sporting to continue when the fly could no longer be clearly distinguished in the failing light. The Cinnamon Sedge (Ronalds) is one of only four caddis imitations in the series. Another, the Welshman's Button, was regarded as an imitation of a caddis species first identified by Halford and Marryat. The "Button" proved saving when the trout refused to come up for mayflies in early season.

A TRAGEDY

In 1907 Halford's wife Florence took her life following a nervous breakdown. That she had been neglected is an unavoidable suspicion. Halford took early retirement from business in 1889 and while he speaks of convivial vacation trips abroad with friends and their wives, he seems to have devoted a great deal of time to his fishing and angling companions. In the *Autobiography* Halford expresses sympathy for the retiree who, never having developed a hobby, finds himself unhappily at loose ends, assuring us that this has never been an issue for him. I'd argue again that his real "business" had been in Hampshire, not London. Take this passage: "For years I had tried to find a stretch of a good stream which I could rent for myself and possibly a few friends, with a cottage or small house in which we could stay and make our home while away on our fishing expeditions." An accomplished person in her own right, Florence apparently felt her individual identity swallowed up by her husband's great fame. She once wrote a clever, rather caustic piece entitled "The Dry Fly Fisherman's Wife" for the *Field*. It was signed "Victim Wife." Halford's passion is documented through five formal books written over a span of about seventeen years not to mention a large number of pieces for *The Field* and *Fishing Gazette*, many under his pen name, "Detached Badger."

Halford fell into a deep depression after the death of his wife. His writing stopped, as did public appearances. Several years passed before his son Ernest and family including two grandchildren joined with William Senior, Martin Mosely, and other friends to encourage him to become active again (*F. M. Halford and the Dry Fly Revolution*). This sadness no doubt delayed the publication of *Modern Development of the Dry Fly*. Halford's lighter comments in the second part of the book, for instance his banter with friends and keepers along the river, suggest that he was recovering.

The Handbook

Halford had several goals in mind when he published his last and largest book in 1910. The fourth edition of *The Dry Fly: Theory and Practice* was out of print, and rather than attempting a revision, Halford preferred to update and to share newer experiences and insights. He felt this necessary regarding insect life and tackle and he wished to respond to the interest his book on fisheries had aroused. While the *Handbook* was not highly acclaimed by critics, I found it useful as a summary of Halford's opinions on all sorts of issues from proprieties to tackle and tactic. In addition, it indicates that Halford's name had become familiar in the retail business. Hardy produced the state-of-the-art Halford Rod with up-locking reel seat in 1912. Aluminum alloy had replaced ebonite for reel construction, pawl click systems were being perfected, and Hardy's Model Perfect reel offered the smooth performance afforded by ball bearings. A Halford Model with an innovative finger release for the spool had come onto the market.

MONTISFONT

In 1905 Halford, by then almost seventy years old, acquired water at Montisfont on the lower Test in Hampshire. Access didn't require long treks, and he welcomed a parade of guests, hoping they would fish his elite patterns exclusively. The congenial flow of angler friends began again. In his last years Halford became a familiar fig-ure around the village of Dunbridge near his fishery, even hosting an annual banquet for the townspeople. Rainbow had been stocked too, much to Halford's delight. The beautiful trout took a dry fly well and fought valiantly. Mr. Valentine Corrie, a close associate, owned a major hatchery operation on the Itchen and earlier had cleansed a lake of pike and eels to create a hatchery for the silvery trout. (Rainbow were stocked in Buckingham Palace waters in 1903, officially welcoming the "Americans" to England.) The rainbow came with a price, though, for grayling were never seen after 1913 at Montisfont.

In his last years Halford's fishing was often restricted to brief eve-nings and he had begun to leave the cold and damp of English win-ters for warmer climes in Spain and France. In February 1914 Halford was suddenly taken with a rapidly progressive and fatal pneumonia while returning by ship from a villa in France. Martin Mosely carried on Halford's work with a new edition of the *Handbook* combined with his *Dry-Fly Entomology*. In his biography Hayter discloses in an endnote that Mosely encouraged Halford to begin some experiments with nymph imitations during the last years of his life. We have no further information.

IN PERSPECTIVE

There's much to like and admire about this complicated man. We've learned how easily he could weave the natural and physical sciences into his fishing, where his level of diligence and compulsive atten-tion to detail clearly went beyond the practical. Clearly he was a natural and gifted teacher. Halford *needed* to instruct, or better, to share. He felt no need to cram his opinions down the throats of his fellow anglers. And if these opinions seemed unduly rigid and authoritative, he had earned certain rights through lengthy study and experimentation.

It's difficult to swallow constant references to "The Dry Fly Man" and the plain assertion that the mantle of good sportsmanship could only be awarded to the ultra purist. We know that some Hampshire fishermen felt that Halford had become especially pretentious when they were advised that his new patterns had superceded their old and proven favorites. I can imagine too that anglers who lacked the financial resources and or social status to access the waters Halford enjoyed would feel envious and resentful. (The extent of Halford's wealth is uncertain; in the *Autobiography* he mentions that there were some limits on his "ways and means.") Halford could be a bit patronizing on occasion as when admonishing the angler to change his damp clothes and that a warm bath "will tend much to the comfort and enjoyment of the evening meal." Mother knows best! But that gets back to "kindly." Halford's fame was well deserved based upon his exceptional personal qualities and accomplishments, furthered and embellished by his access to the media, all coming together at a propitious period in time. As far as acknowledging a debt to the great anglers who had gone before, Hayter said, "He seems to have had little curiosity about the past, and almost never quotes from the old angling authors except Ronalds and Frances Francis. In some respects he resembled the men of the renaissance, who were carried away by the achievement of their own age and despised their medieval predecessors, doubting if anything valuable could have been achieved by them" (F. M. *Halford and the Dry Fly Revolution*). *Modern Development of the Dry Fly* was intended to be Halford's last book and hence these parting words:

> Many of my readers are my friends; some few have dealt with my theories in a critical, but I hope not hostile spirit; many strangers have corresponded with me at various times on matters connected with our favorite sport. So far as I know, there is not one of them who would call himself my enemie. To one and all let me make my farewell bow, and thank them for their praise, their criticism,

their friendship and their correspondence. They have all contributed their share to the interest and pleasure of a comparatively long and happy life.

CROSSED RODS

With apologies for redundancy, I conclude this chapter with a brief summary of Halford's positions as an introduction to Skues. The relationship between the two men began most cordially. Indeed Halford and William Senior sponsored Skues for membership in the Fly Fisher's Club in 1893 and the two later served together on a committee. The battle of words began in 1897, when Skues wrote a rather critical review of *Dry-Fly Fishing Theory and Practice*. Halford considered it less than respectful. Both men were frequent journal contributors, and so the friction grew. Hayter mentions an unflattering review of *Modern Development of the Dry Fly in the Fishing Gazette* where Skues wonders how the elite patterns could have been tested if they were not fished against standard patterns. As Hayter said, Skues would never have been a guest at Montisfont. Those irritatingly truculent bulgers, "ploughing" about beneath the surface before or during a hatch were the central issue in the conflict with Skues. Halford stated repeatedly that bulgers should be left alone because experience had shown that they would not rise for a floating fly. Hence the dry-fly man would be wasting his time. Secondly, pestering these trout might make them shy in the event that they eventually renounced bulging in favor of feeding properly on the surface. (Whether the man from Yorkshire had success with bulgers is not stated.) When others, primarily Skues, began to write of catching bulgers fishing upstream and "wet," Halford felt it necessary to respond in the *Handbook*. "Flies are either dry or wet; there's no half-way house. I am told, however, that there is a school of fly fishermen who only fish the sunk fly over a feeding fish or one in position if it will not take a floating fly. This they urge, is a third method of wet-fly fishing, the other two being the more ordinary *fishing the water* with sunk fly either upstream or downstream. Candidly I have never seen this method in practice and I have grave doubts as to its efficacy." Those were fighting words. The school to which he referred was still pretty much a one-man school and Skues was its headmaster.

Sir Edward Grey and
the Interregnum

*H*alford achieved the pinnacle of his ascendancy following the publication of *Dry-fly Fishing in Theory and Practice* in 1889, some twenty years before Skues introduced his new ideas in *Minor Tactics of the Chalk Stream* (1910). You wonder how the beleaguered wet fly fared during that span. How would a fair-minded chalk-stream angler, especially one outside Halford's circle, view the wet fly, now relegated to the chimney corner? Edward Grey's classic *Fly-Fishing*, published in 1899, provides such a balanced view and further helps to explain why the nymph lagged behind the dry fly in both theory and practice. Hills called Grey "the best and most devoted dry fly fisherman in England and gifted with the power to write fine prose" (*A History of Fly Fishing For Trout*). (Where did that leave Halford?) Hill's accolade might suggest a dry-fly bias on the part of Grey, but in fact Grey maintained a lifelong love for the wet fly and is credited with calling attention to the sea trout as a worthy quarry for the fly fisherman. Grey is an important figure in history quite apart from angling.

From Winchester to Parliament

The Greys could trace their family lineage back several generations to an early prime minister. Their territorial designation was Fallodon, a village in Northumberland. Edward fished regional streams and in

Scotland as a boy before entering the prestigious Winchester College (a public school, or what Americans would call a prep school) in 1877. The Itchen runs through the ancient cathedral city and to parody Norman MacLean's *A River Runs Through It*, the legendary chalk stream runs thorough the very core of fly fishing's history. Sir Izaak Walton is buried in Winchester Cathedral, and the college's founder, William Wykeham, is said to have been the first to experiment with floating flies way back in Dame Juliana's day. The college's curriculum, however, failed to take advantage of its campus river. Indeed young Edward was one of only a few boys who took interest. Told that there was little chance of success he nonetheless dashed, rod in hand, back and forth between the river and his classes or meals. The difficult Old Barge water where he watched Marryat work his magic became a favorite, although he managed to catch just one trout during his first year. During the 1880 season he recorded seventy-six fish taken on dry flies, the result of his dedication and of lessons learned by watching others. (Halford and the much younger Grey were both honing their skills during this period, at times presumably, on the same river.) After graduation Grey went on to Oxford. At about this time, following the death of his grandfather, he became *Sir* Edward and, coincidentally, expert in tennis, later winning the British Empire Championship five times. He was married in 1885 and the same year became the youngest member of the House of Commons. After a series of lesser diplomatic appointments, Gladstone named Sir Edward Undersecretary of State for Foreign Affairs in 1892. Three years later Grey assumed the senior role for an unprecedented period of service from 1905 to 1916. Anyone with an interest in the Great War will find that Grey was a major diplomatic figure, engaged in one international crisis after another as hostilities threatened. Among other achievements he was responsible in part for the strengthening of relations with France and Russia known as the Triple Entente. When diplomacy finally failed, Grey made the sober comment "The lamps are going out all over Europe and I doubt we will see them go on again in our lifetime." In 1916, he was granted a peerage (Viscount Grey of Fallodon) and as a member of the House of Lords became the President of the League of Nations. Grey was later appointed Special Ambassador to the United States. Despite failing eyesight, Sir Edward was considered a master ornithologist (an institute is named in his honor) and his book *Charm of*

Birds was highly popular. He and Colonel Teddy Roosevelt became close friends through their studies comparing English with American bird species. Grey wrote several books concerning the war, eventually retiring to Fallodon where he died in 1933.

Regarding Halford

In his book Grey treats Halford with a sort of distant respect, giving no indication that they were well acquainted. He agrees that it is best to wait for the rise to begin and to work to a particular fish in position. Like Halford, he disrespects bulgers. When trout are rushing about after larvae "it is better to be casting over the most fastidious trout which is taking flies on the surface, than over the hungriest one that is bulging." Both admired a well-cocked, high-floating fly. "In fact the fly must float as if it were buoyant, cheerful and in the best of spirits—natural flies having the appearance of being very frivolous and light hearted." Although small flies and fine gut were advised, Grey did not believe in imitation as such, employing just four patterns. Perhaps following Halford's lead, Grey experimented with hatchery culture, attempting to raise rainbow and brook trout in addition to browns. His critical opinion of the Mayfly was hardly in accord with the adulation typically accorded the big insect. "May is a good month on a chalk stream, but to my mind the perfection of dry fly fishing is to be had on a good day in mid-June, on water where the May-fly never appears, first to excite the trout and the anglers, and then to leave the fish without appetite and the angler too often discontented." Grey felt that an equable continuation of a series of lesser mayfly hatches through June into July with correspondingly steady sport was more likely in streams uninhabited by the Mayfly. Overstocking was to be avoided, and Grey welcomed enough pressure to keep the trout educated and alert. *Fly Fishing* is one of those happy books that instructs informally through the recounting of illuminating experiences that will hopefully be familiar to the reader. Separate chapters treat of the dry fly, wet fly, sea trout, salmon, tackle, and stocking.

The Wet Fly

The enthusiasm which was the result of dry fly fishing led at one time, among those who were fortunate enough to be able to enjoy it, to a tendency to disparage the older art. A comparison of the two methods is always interesting, but it must never be forgotten that it is not necessary, nor even appropriate to exalt one at the expense of the other (*Fly Fishing*). I liked that. Surely it was fully understood that the dry fly was the upper-crust way, but no one had laid out the fact so plainly nor discouraged resulting value judgments. Grey admits that the dry fly "possesses the South of England" while the wet fly holds sway in the "North and in Scotland," presumably on the basis of different types of fishery. He then alludes to the debate in the chalk streams.

> Some dry fly anglers may have spoken of wet fly fishing as a "chuck and chance it" style. Some wet fly anglers may have expressed a belief that all the talk about dry flies is superfluous, and that large well-fed trout in clear smooth water can be caught by the methods skillfully applied, which are successful in north country waters. If there be any angler on either side, who still holds such opinions, he can but be advised to put them to the test in practice, and so bring himself to a more just frame of mind. But controversy is not always the best method of deciding what is the truth, and in most matters connected with angling, partisanship leads to error, just as certainly as in other affairs.

That's diplomatic—and fair! Grey then pursues a different tack by suggesting that the chalk stream has been a perfect laboratory where "the habits, both of the trout and the insects upon which they feed can be studied almost as in an aquarium." The reference to Marryat, Halford, and others is clear, but interestingly, Grey indicates that these advances have been restricted to the dry fly. Upstream versus down was necessarily part of the discussion, and here Grey reveals his admiration for Stewart. "Much has been written about the proper method of fishing wet flies, whether it is best to fish upstream of downstream. It is easier to argue in favour of the upstream method, and if two men of equal ability held briefs on each side, and argued the case against each other before a jury who were without experience, of either method and therefore presumably impartial, the verdict would probably be given for fishing upsteam." Grey nonetheless appreciated the difficulties in reacting promptly to a take on a slack line. "At the end of a day's fishing we know of fish that have been touched or risen, but who can know how many have taken he fly and rejected it, unfelt and unseen? The probability is that any trout, which takes the fly, will not be hooked or even felt when the line is slack." For this reason Grey favored the downstream approach in general. He could not have anticipated that Skues was to utilize the pellucid chalk stream in the study of the trout's reactions to hatching mayfly larvae and would teach us to overcome the natural drift that is the challenge of setting the hook upstream. Those insights were to come ten years later.

Sea Trout

Grey stands alone in singling out the sea-run brown trout as a distinctive prize for the observant angler. Knowing nothing about the fish, I found his presentation quite compelling. As the trout season began to wear thin in late July, Grey sensed that the "gentle and docile" chalk streams were becoming soft and effete and yearned for the more exhilarating and rugged sea-coast rivers. Late summer was also not the best season for prize salmon. Accordingly, this period of five or six weeks needed a spark—enter the sea trout. Sometimes I think that sea-trout fishing is the best of all sport. It combines all the wildness of salmon fishing with the independence of trout fishing, and one may have all the excitement of hooking large fish without using a heavy rod and heavy tackle. There is less formality about it. Grey understood why, sharing the water with much larger, active (but seldom catchable) salmon, the sea trout had been overlooked. And the fish, silvery and with a forked tail and different markings, resembled small salmon far more than brown trout. (Grey believed them to be separate species.)

The best and least expensive sport was found along smaller rivers that drained lochs. Here the fish sought out deeper, quiet stretches and took best when the water was falling after heavy rains. The same patterns suited for salmon were sufficient, presented in the same way. It was nonetheless difficult to guess just where the sea trout would hold, so experience with each river paid off. The sea trout's feeding habits were capricious, ranging from refusals to diffident rises to smashing takes that snapped heavy gut. Overall they took much more readily than the salmon, and Grey warned against wasting time casting to them when sea trout in the one-to-four-pound range were a much surer bet. The sea trout is a wild and mysterious creature without a home, and its habits differ as much from those of brown trout as the habits of wild fowl or woodcocks do from those of partridges.

I prefer a good fresh-run three-or-four-pound sea trout in a river on a single-handed rod and fine tackle to anything else. Of all fresh water British fish, the sea trout fights the best in proportion to its size. Today lodges and fly shops in England often mention sea trout as an option along with salmon and fresh-water browns and even offer instruction specific to the sea trout. Elsewhere, as in Patagonia, there are few fly-fishing adventures as exciting as the pursuit of sea-run browns.

Sir Edward's Place

He is remembered largely in the UK, largely as a statesman and a popular ornithologist. *Fly Fishing*, sandwiched between Halford and Skues, was his only angling book I nonetheless highly value Grey as a writer whose particular sensitivity allows him to share his great love for the sport. Here he deals with one of life's conflicts. However Halford may or may not have struggled with the conflict between the clothing business and his fishing, the issue was far more poignant for Grey.

> This is not the place in which to write of the deep human interests of London, of what great affairs have their centre and of what issues are discussed and decided there. But there is an aspect of London which is inevitable and becomes most oppressive on hot June days. There is the aggressive stiffness of the buildings, the brutal hardness of the pavement, the smell of the streets festering in the sun, the glare of the light all day striking upon hard substances, and the stuffiness of the heat from which there is no relief at night—for no coolness comes with the evening air, and bedroom windows seem to open into ovens. It is impossible to live in London without great sacrifice.

Grey goes on to recommend living within walking distance of Waterloo Station where the 6 AM train to Winchester awaits. Within an hour "you step out of the train among all the long-desired things. Every sense is alert and every scent and everything seen or heard is noted with delight. You are grateful for the grass on which you walk, even for the country dust about your feet." In his concluding chapter, Grey speaks of the days of his youth spent fishing the north-country burns "too small for fly fishing, which run in narrow stony channels between overgrown banks." Here he learned stealth, offering a

Brandling worm on hands and knees, even as Walton. Rather than simply reminiscing, Grey is serious about burn fishing, tempting the reader, presumably an adult, who might enjoy the primitive yet difficult art. "There are times when I have stood still for the joy of it all, on my way through the wild freedom of a Highland moor, and felt the wind, and looked upon the mountains and the water and the light and sky, till I felt conscious only of the strength of a mighty current of life, which swept away all consciousness of self, and made me part of all that I beheld." I rest my case.

G. E. M. Skues Picks
Up the Gauntlet

*R*egarding George Edward Mackenzie Skues, there's an important matter to be dealt with up front. His pronunciation of his name—"Skew eez"—sounds like a violinist's embarrassment or maybe a vehicle with loose power steering belts turning a corner. It's not important unless you assume it's "Sk use" when trying to impress other folks who know better. (I'm indebted to Glen Law for setting me straight.) More important than the sound of his name, Skues built a large part of the foundation on which fly fishing rests as we know it today.

Winchester—Again

Born in 1858, Skues was fourteen years Halford's junior. His father and mother were often abroad, engaged in government service, leaving him to live as a child with his grandparents. Later Skues grew up a boarding school boy. Perhaps being largely on his own at such an early age helped him develop the independent nature that led him to later question Halford's dogma. Entering his teens, Skues won a competitive scholarship to Winchester College. Marryat, Halford's great friend Tom Sanctuary, and Edward Gray were only a few of the many gentlemen anglers who undertook to learn the art on the school's storied waters. According to dates, Skues's last year at the school would have been Grey's first, but to my knowledge, neither man mentioned the other. John Hammond's shop in the city was the source of the best in rods, flies, and advice, serving as a gathering place and nexus for communication among the elite chalk-stream men of the day. With the Itchen close at hand, Master Skues longed to learn the popular dry-fly method despite lacking funds for decent tackle and absent an understanding of the intertwined biology of trout and mayfly. Robson's account suggests that Skues's mastery of the dry fly probably fell well short of Grey's (*The Essential G. E. M. Skues*).

Out of curiosity I looked up the Winchester College website, finding a list of notable graduates from each century, going back six hundred years. Nineteenth-century alumni included Lord Grey; Anthony Trollope; Arnold Toynbee; George Mallory of Everest fame; Lord Alfred Douglas, poet and companion to Oscar Wilde; and John Montague, suspected of being Jack the Ripper. Unaccountably, Skues and Marryat didn't make the honor roll. That committee was badly in need of a fly fisherman or two! (Amazingly, the Cathedral boasts of Jane Austen's burial site with no mention of Walton's.)

A HALFORD DISCIPLE

Skues entered a law clerkship shortly following graduation. It was a fortuitous choice, for in 1883 he met Irwin B. Cox, a client of his firm, part owner of *The Field* and a chalk-stream habitué. Cox took a liking to the eager Skues, inviting him for a day on the Itchen, where he was fortunate to meet the influential Francis Francis. It was an opportunity for a young angler to penetrate the fringes of the Hampshire nobility. Skues was later given carte blanche to fish Cox's beat. (Cox later became one of Halford's closest friends.) Now well equipped and better informed, Skues began to visit fisheries throughout England and others in Europe. After receiving a copy of *Dry Fly Fishing, Theory and Practice* in 1887 he began to tie flies, and two years later, as Robson puts it, the book had become "a sort of gospel." In 1891 Skues finally met Halford on the Itchen. Whether because of his prowess or persona or both, Skues was thereafter accepted into the chalk-stream pantheon and invited to join the prestigious Fly Fisher's Club. Robson makes excellent use of Skues's letters and those of others relevant to a particular topic. Intriguingly, during days shared with Halford on the water, presumably fishing dry flies, Skues tells us that he consistently "wiped his [Halford's] eye."

Undisturbed, Halford congratulated the younger angler on his success. Rather than disrespecting Halford or exalting his own prowess, these "victories" merely encouraged Skues to question dogma and to learn from his own experiences (*The Essential G. E. M. Skues*).

THE MATCH IS JOINED

Halford's sour remark about the "third wet fly method" collided in 1910 with the appearance of Skues's first book, *Minor Tactics of the Chalk Stream*. Halford had learned about the distasteful practice of fishing nymphs upstream to bulging trout through a piece Skues wrote for the *Field* in 1899, suggesting that the technique might make sense when feeding trout were not feeding on the surface. *Minor Tactics* further developed, explained, and defended Skues's thesis. His title is modest if not self-effacing, and the dedication is curiously conciliatory: "To my friend the dry fly purist and to my enemies, if I have any." The foreword too is gently crafted with a bow to Halford: "I have been led into a sustained effort to recover for this generation, and to transmute into forms suited to the modern conditions of the sport on the chalk-stream, the old wet-fly art, to be used as a supplement to, and in no sense to supplant or rival the beautiful art of which Mr. F. M. Halford is the prophet." Skues's tenor thereafter becomes openly declarative and is not without teeth, if only politely bared. From the first chapter:

> In no art are its practitioners more slavishly content, than in angling. Tradition and authority are so much and individual observation and experimentation so little. In all humility is this written for I look back upon many years when it was my sole ambition to follow in the steps of the masters of chalk-stream angling, and to do what was laid down for me and no other; and I look back with some shame at the slowness to take a hint from experience which has marked my angling career.

It should be understood that Skues always looked forward to those days and times when the trout were taking well on the surface, for he was as much a dry fly enthusiast as any of his colleagues. That is exactly what he intended to convey in "*Minor Tactics*." For the sake of clarity, Skues's "Tactic" might be paraphrased: "When expertly presented dry flies are refused by trout that are presumably feeding, offer them a nymph, drifting naturally like a dry fly, close to the surface."

Much of the contents of *Minor Tactics* is given over to practical issues equally applicable to either the wet or dry fly. There are suggestions regarding the choice of a bank as influenced by the wind, bank structure, and where the trout are holding. The feeding habits of the trout are a recurring theme, and there is an overview of effective casts. One section offers advice about playing large fish and the wisdom of letting them run under gentle pressure. Skues suggests a nifty trick for inducing a "weeded" trout to come out of its stronghold and throughout the text is brightened by intriguing accounts of the killing of difficult trout under unusual circumstances. *Minor Tactics*, then, is as much a chalk-stream primer as an advertisement for an innovative technique. Terry Lawton makes an important point in that *Minor Tactics*, like all Skues's books, was a collection of pieces published earlier in a variety of sporting journals, collated as to topic and stitched together with a foreword and introduction (*Nymph Fishing: A History of the Art and Practice*, 2005). This is significant in that otherwise one might assume that readers only became aware of what Skues had to say after a particular book appeared.

Early Experiences

The anecdotes that showed the appeal of a shallowly submerged fly drifting with the current were collected over time. One sweltering, humid day the trout were "smutting" or gorging on midges caught in the surface film. They refused dry flies, as expected with smutting fish. By chance Skues tried a Dark Olive Quill, well tied, but with hen's hackle, and when he could no longer keep it afloat, the trout ate it up. He found the trout's stomachs packed with midges, but no mayfly parts, so this was not a "masking hatch." The experience was put aside as preternatural. On another occasion, several years later, Skues was frustrated by a heavy and sustained mayfly hatch, the trout "lined up in force to deal with them and feeding steadily." Picking out some of the largest fish, Skues covered them carefully, trying an assortment of nine patterns without putting them down—and also without success. After a lunch break he gave it one more try, this time with an old-fashioned dry-fly pattern, also tied with hen hackle. Just as soon as it soaked up and sank, he began to catch fish after fish. He called this "another hint as pointed and definite as one could get from the hind leg of a mule. Still I did not realize that I was on the edge of an adventure, nor yet where I was tending when Mr. F. M. Halford told me how a well-known Yorkshire angler had been fishing with him on the Test. . . ." We know the rest of that story, although not just when Halford made this disclosure.

Dry Flies That Became Nymphs

The Tups Indispensible [*sic*] was created in 1900 by one of Skues's many correspondents, R. S. Austin of Devon. It was intended to be fished dry as an imitation of a red spinner, but Skues "found it more successful semi-submerged than floating and by no means a bad representation of one of the nymphs." The pattern was tied in pastels, the translucent body with a light yellow butt giving way to light pink behind the ginger-tinged wings and pale hackle. The Tup's (a dry fly) was so popular that Austin tired of tying so many dozens, but nevertheless kept the dubbing materials secret. Sworn to secrecy, Austin shared the recipe with Skues, who gave the pattern its curious name. Who was Tup? It turns out that the fine yellow wool that covers a ram's testicles was the magical element. When a ram mounted a ewe he was "up" or a "tup." (How would you go about shaving a ram's privates?)

Skues began serious efforts to modify various dry-fly patterns to more resemble the naturals, for example removing the wings from a Tup's Indispensible, and to make them sink more readily. He also began to collect and copy natural larvae (*Minor Tactics*). Commercial "nymphs," grotesquely overdressed, were held in the greatest contempt. He set forth armed with a small butterfly net attached to the handle of his landing net to sample the film and subsurface during mayfly hatches. Better known, his famous "marrow spoon," a narrow, thin-bladed implement, was handy for scraping out the contents of gullets. The remains were then washed and teased apart in his white porcelain "baby dish." The marrow spoon eventually caught on with serious nymph fishermen, even in America. To Skues's astonishment his spoon once recovered baby toads from the throats of some greedily feeding trout; a large dry fly, chopped down to shape, turned the trick. (Probably the only reason Skues didn't come up with the suction bulb technique is that his fish had, mercifully, been killed.) Although never in Halford's class, Skues was a capable, hands-on entomologist and also turned to Martin Mosely for advice in correlating larvae with the corresponding adult insects. (Although the relationship between Halford and Skues grew ever distant, they shared many friends, such as Francis and Mosely.) To his great satisfaction, imitations of the Dark Olive and Blue-Winged Olive larvae were readily accepted. A particular circumstance brought additional admiration for the nymph. A prolonged period of drought left the rivers low and warm by midsummer. There were few duns or spinners in evidence yet trout rose fairly actively. When dry flies brought no response, Skues's nymphs were welcomed. Small mayfly larvae made up 95 percent of stomach contents (*Minor Tactics*)!

Those Bulgers

They were another matter. In *Minor Tactics* under the heading, "Of Medicine for Bulgers," Skues quotes another angler to the effect that "when trout are bulging, you might as well throw your hat at them." For some time that was also his experience. Hare's Ears and many other patterns were refused until one day he finally caught a brace of bulgers at the beginning of a hatch on a Pope's Green Nondescript, a pattern imitating a dun. Why would the trout mistake a winged fly intended to represent a dun or sedge for a nymph? He wondered if the trout, looking upward, react to the reflected image of a sunken fly in the surface film? (I assume Skues thought that the wing's contribution to the image would then be altered, possibly suggesting erupting wings?)

Fishing the Tweed, he noted that the trout were nearly always bulging and readily accepted the local dressing of the popular Greenwell's Glory, long a mainstay, tied both wet and dry. Greenwell, a canon of the church, developed the pattern on the Tweed in the mid-nineteenth century during a March Brown hatch. It sported a ribbed body of waxed yellow silk, a thin, swept-back wing of blackbird or starling, and brown, beard hackle. The wetted body afforded translucency. The Greenwell's was never popular in America, but famed angler Vincent Marinaro liked its looks and tied several as floaters. They didn't! "I began to regret the waste of labor and material expended on those useless dry flies that I could not float even with the most vigourous and frequent false casts." Then a hatch began and Marinaro "was getting a strike on almost every pitch. There was no doubt at all that these fish were taking emerging duns" (*In the Ring of the Rise*, 1976). When Skues went on to combine the colors of the Pope's with the narrow wing style of the Greenwell's, the result was an effective solution to the "bulger problem." He proclaimed, "I have provided myself with a method which forms an admirable supplement to the dry fly."

Skues was not a stickler for matching size and on occasion fished a fly a bit larger than the natural. He wanted his fly to sink promptly, and to this end he used heavier hooks, omitted tails, thinned bodies, reduced the quantity of hackle (short and soft-fibered), and bulked up wings so they would sop up more water. (Glycerin was occasionally applied to a fly when it was to sink immediately even though the fly and tip of the leader were never more than a few inches below the surface.) Fine gold ribbing served to add a bit of flash. Tails were either sparse whisks or omitted altogether. Seal fur dubbing was strongly preferred because of its brilliance and translucency. The feather wing case, tied in near mid-shaft, was drawn forward to the eye and tied off or the tips might be folded back again to protrude as a whisk of stubby legs. Skues vigorously addressed the ageless issue of whether trout see as we do, deciding finally that they do not, and yet he believed color to be a critical component of imitation, more so than form or silhouette. Accordingly, he was frustrated when the trout eagerly attacked a pattern during a hatch when the fly's coloration was distinctly unlike that of the natural insect. For instance the Gold-Ribbed Hare's Ear, fished dry, was a favorite of both Halford and Skues, doing great harm to the trout when large olive duns were on the water. The brownish, shaggy fly little resembled the smooth olive-green-bodied natural. Despite their confidence in the Hare's Ear, when no acceptable explanation for its success could be found, both men eventually stopped using it. Skues bent this ethic in the case of his most dependable Orange Quill, coloration nothing like the naturals. He rationalized that trout see what they want to see—if the rest of the fly looks inviting, the hook is ignored, or maybe taken for a shuck. He summed things up with a bit of a sneer: "The imitation may be Impressionist, Cubist, Futurist, Post-Impressionist, Pre-Raphaelite, or caricature. The commonest is caricature. It therefore catches the most fish" (*The Way of a Trout with a Fly*, 1921).

About Drag

"Of all trials of the chalk stream angler, perhaps drag is the worst" (*Minor Methods*). There was general agreement that the dry fly, drifting slightly off speed or direction, would almost always be refused or put the fish down. However the trout seemed to tolerate, even appreciate, a wet fly's drift, spiced with a bit of drag. (This idea was a cardinal and consistent point of attack against the unsportsmanlike wet fly.) Later we'll meet a number of American anglers adept at introducing slack into their casts to forestall drag. It's hard to imagine that Skues lacked these skills, but chalk-stream currents were less vigorous than those of our swift streams, and he emphasized precise presentations that dropped the fly just far enough ahead for a fish to see it and react before drag could set in. Bulgers were a problem. "Bulging trout are bold feeders, and seem to mind being cast over less than do those which are taking surface food; but they are much more difficult to cover accurately because they rush from side to side and up and down, and the odds are that, if you cast your fly to one spot, the trout is careening off in pursuit of a nymph to the right or left of it" (*Minor Tactics*). Skues gives the impression that during a busy emergence the trout dart about in competition with their fellows and are seldom discouraged by inadvertent drag. Occasionally he observed a fish holding and feeding in mid-water. Skues liked to drift a nymph by its nose, a departure from the usual approach. Of note, the fly was sometimes taken when the rod was lifted for a fresh cast, thus swimming the fly toward the surface (*The Way of a Trout with a Fly*). (Although this is mentioned merely as an aside, in recounting the experience, Skues anticipated, unknowingly, the next great advance in nymph presentation.)

How to Know?

The difficult issue of how to detect the take of a submerged, drifting fly had to be addressed. Here Skues's lucid observations are more practically descriptive than any I've read since. He talks mostly about subtle signals such as flashes, shadows, and a bulge on a glassy surface film. Skues's delightful verse concerning the "cunning brown wink under water" is still quoted; this is only the first of eight stanzas from the *Little Brown Wink*, originally published in the club's journal, *The Way of a Trout with a Fly.*

Oh, thrilling the rise at the lure that is dry,
When the slow trout comes up to the slaughter,
Yet rather would I
Have the turn at my fly,
The cunning brown wink under water.
The cute little wink under water,
Mysterious wink under water,
Delightful to ply,
The subaqueous fly
And watch for the wink under water.

The downstream wet fly angler, timidly contemplating fishing his flies upstream, must surely have found enlightenment and encouragement in these few pages. (The practice of "oiling" the leader as an indicator was later attributed to Skues's brother Charles; it's not clear whether Skues ever used the assist.) Having informed and reassured the reader, Skues now goes on to elevate the accomplishment to something of a religious experience:

The indications which tell your dry-fly angler when to strike are painfully obvious, but those which bid a wet-fly man raise his rod-point and draw in the steel are frequently so subtle, so evanescent and impalpable to the senses that, when the bending rod assures him that he has divined aright, he feels an ecstasy as though he had performed a miracle each time. In an effort to divine the indications which call for striking with the wet fly I confess I find a subtle fascination and charm, and, when success attends me, a satisfaction beside which successful hooking of a fish which rises to my floating fly seems second-rate in its sameness and comparative obviousness and monotony of achievement.

Those sentiments must have made the dry-fly man's hackles stand on end. Yet with all this Skues only claims supremacy for his method under conditions when dry flies are typically refused, readily admitting that "there are days and times when the dry fly will beat the wet fly all hollow." Given the necessity of reacting quickly to any signs of a take, he was curious as to how trout engulf or discharge a fly. Experiments were therefore conducted in the bathtub. The image of the portly Skues semi-submerged is rather droll. While he was unsuccessful in re-creating the bubble a trout leaves when taking an insect off the surface, he did learn that trout draw insects into their mouths through suction rather than biting and blow rather than spit them out (*The Way of a Trout with a Fly*).

Skues and the "Emerger"

The events attending the "birthing" of an aquatic insect have great significance for the fisherman, wet or dry. This is true to the extent that we have invented our own word, *emerger*. You won't find it in a dictionary. According to Robson, "The combination of low-floating dry flies, so called emergers or nymphs fished in the surface or fractionally below it, and in the upstream style to rising fish owes its origin to Skues" (*The Essential G. E. M. Skues*). How did he justify this accolade? First off, Skues understood the particulars of emergence in some detail. This description comes after observing a heavy hatch on the Kennet:

> The dull, inert, brownish body of the nymph, swung down by the current from the swaying tassels of water weed, coming slowly to the surface, till somehow its head and thorax seemed to threaten to emerge. Then the bursting of the brown skin of the thorax, the six pale green legs gripping the surface, while the body curled tail downwards in the water, as if to let the current get a purchase on the sheath. Then simultaneously the wings shooting up, the sheath coming away and floating far down the current, and the fairy-like creature standing with wings erect and upturned tail to drift down-stream, it may be a few yards or only a few inches before taking flight for the meadows. (*The Way of a Trout with a Fly*)

Nymphs floating with the current, wings not yet erupted and enclosed in a bubble of air or gas beneath the envelope of the outer skin were also described.

I cant resist this paragraph as an example of Skues's style. It's entitled "The fumbling of it," from *Side-lines, Side-lights and Reflections*.

> The up-winged dun when it is once hatched out, is such a lovely, delicate, clean-cut little creature that one is apt to forget that at the moment of eclosion when it emerges from its envelope which clad in its nymphal form, it passes through a stage of untidy struggle not distantly resembling that which a golfer or footballer displays

in extricating himself from a tight fitting pullover or sweater or jersey. Yet from the fly fisher's point of view this is a very important stage, for it is at this that the trout, when not confining himself to duns on the surface—and he seldom confines himself strictly and exclusively to this—must be constantly seeing and absorbing the insect. Sometimes, indeed frequently, the operation of emergence is performed with neatness and dispatch. Often however, it is an awkward and fumbling occasion, affording the fish a much longer and better chance than is given by the dun which hatches neatly and quickly. It may also be that the hackled pattern, dressed with soft, henny fibers closing in on the body, may afford no bad suggestion of a subimago struggling, as it often does, with crumpled wing to emerge from its envelope. I thus satisfied myself that Tup's Indispensible could be used as a wet fly; and, indeed, when soaked its colours merge and blend so beautifully that it is hardly singular; and it was a remarkable imitation of a nymph I got from a trout's mouth. The next step was to try it on bulging fish, and to my great delight I found it even more attractive than Greenwell's Glory.

As time went on, Skues preferred to fish his Tups "semi-submerged" in the film, essentially a floating nymph, conforming to his descriptions of the naturals. There was more to the pattern than the ram's contribution. Skues described some combination of cream seal's fur or mohair and "filmy fur from an English hare's poll." He tied a series of Tup's nymphs by varying the quantity of seal's fur and color of the mohair. The dry fly's cream and bluish hackle was replaced with short hen's-hackle fibers. Skues concludes, referring to emergence: "It is a fact that nowhere in any angling book does one find a direct suggestion that the natural fly should be imitated at this stage." Skues also suspected that the image of an emerger trapped in the film and that of a spinner might appear similar to the trout. He fished a Tup's to good effect when spinners were adrift, further suggesting that Halford's spinner imitations were sometimes taken for floating nymphs! The caddis emergence was thought to be too volatile for successful imitation although Skues noted that a winged dropper fly, if twitched in the skin of the surface, might suggest "something alive and in difficulties." (An aside: Famous old patterns tend to change their original costumes almost beyond recognition over time. Google Tup's Indispensible and you'll find, among others, a long-hackled spider up in Yorkshire, a parachute treatment, and a brown hackled version. Bodies range from all yellow to 90 percent pink, bead-headed or otherwise, and so on. Mr. Austin would have been appalled.)

THE WIGGLE ISSUE

Halford and Marryat never questioned that nymphs were an essential food form for the trout and agreed that bulgers were probably feeding on them. They had further observed that larvae in and over shallow weed beds were active creatures, darting and squirming about, and concluded that the same aquabatics must continue in the top water up to the actual emergence. The crux was their belief that these antics would be essential and impossible to reproduce with immobile artificials. As Marryat famously remarked: "You can imitate the nymph, but you can not imitate the wiggle" (*F. M. Halford and the Dry Fly Revolution*). This wiggle factor was typical of Halford. When what he thought to have been a thorough and thoughtful exploration of an issue led to a firm conclusion, he considered the matter closed and final. There were no second thoughts or disquieting insecurities in the event of conflicting evidence.

Skues also observed larvae capriciously cavorting above and through the fronds of weed beds and had entertained himself by catching them with a dragged fly. But here came an absolute parting of opinions. From *Minor Tactics:* "The nymph on the point of hatching is not the active, wriggling creature one extracts from sand or weed or mud, but a more or less inert and comatose being, tossed hither and thither by the current and apt to drift out of the strength of the current into the bays and eddies under the banks where trout lay in wait."

How to resolve the wiggle debate? My interpretation: Skues never indicated *where* in the water column the active streambed larvae near the streambed presumably become quiescent. There was no evidence that he believed the insects of interest to the bulgers were *necessarily* immobile. As we've read, it was when the insects, fully committed to hatch, neared or reached the film that their antics diminished or ceased. (Admittedly, Skues's description of the "untidy struggle" from the preceding paragraph doesn't exactly jibe with "inert and comatose.") In any case a number of writers have suggested that the pliant, motile hackles of his nymphs likely simulated the wiggle of a hatchling escaping the shuck. Observations of pre-emergent larvae in aquaria later proved that the larvae of many insects ascending the water column are dynamic swimmers—score one for Halford! But what really mattered was what happened in the river—score a bunch for Skues!

Bottom Water, Off Limits!

Recall that Cotton rather disrespected the fish that live near the bottom, and likely the anglers who fished for them as well. Marryat, Halford, and Skues shared the notion that fishing deep water was unsportsmanlike. It appears there was an analogy to using bait. Skues once experimented with an imitation of alderfly pupae recovered from a muddy bank. The results were impressive, but he felt qualms "as to whether it was quite the game to imitate this insect at this stage." He sheepishly admits to fishing his version of a caddis worm: "I enjoyed mad success for a few minutes and gave up, conscience stricken." He also fashioned an imitation of a "shrimp" with a shell-like back, but unwilling to try it, gave the fly to a friend who promptly lost it to a very large fish (*Minor Tactics*). Skues reiterated his conviction that streambed larvae, the stonefly, caddis, and midge are safe from the wiles of the fly fisherman.

I've found only one mention of weighting a fly or leader by any author up to this period and can only imagine that using weight (in whatever form) sufficient to bring the fly to bottom was simply regarded as unsporting. The idea is too obvious otherwise. Skues's close friend Dr. E. A. Barton fished with him on the Test and kept a diary of their experiences. At least on one occasion Barton fished a nymph weighted with fuse wire wrapped around the shank. Robson notes that there is no indication of Skues's reaction although he must have been aware of this unusual ploy.

Skues tied a representation of the important early-season Grannom caddis pupa with a green silk or wool body and brown partridge hackle. It's curious that fishing a copy of the caddis "worm" was apparently shady practice whereas imitating the pupa was marginally more acceptable, at least in the north. The slow-flowing depths of the chalk streams proved difficult to probe. Skues even wrote, "A trout that is glued to the bottom is generally a pretty hopeless fish." In this he was to be proven wrong not many years later.

A Large Burr Under the Saddle

In the case of Halford, Skues, and others who produced a series of publications, I like to read them in sequence. Skues's second book, *The Way of a Trout with a Fly*, followed *Minor Tactics* by a decade. I wondered in particular how the "Halford matter" would be addressed, posthumously so to speak. The clever title is a parody of Proverbs XXX: "The way of an eagle in the sky, The way of a serpent on a rock, The way of a ship in the middle of the sea, and The way of a man with a maid." You could see a way to work nymphs into that, and Skues's preface follows suit: "Our aviators seem to be on their way towards a solution of the way of the eagle in the air. The mystery of the way of a ship in the midst of the sea has yielded all of its secrets to the persistence of modern man, but the way of a man with a maid and the way of a trout with a fly remain with us to be a delight and a torment to thousands of generations yet unborn." Skues certainly had a wry talent, tongue never far from cheek. The dedication to the book is as placid and peaceful as any chalk stream: *To the Fly-Fisher's Club in Gratitude for Many Happy Hours and Some Priceless Friends.*

I didn't anticipate any trouble. In *Minor Tactics* the dry fly was certainly given its due, and aside from his suggestion that setting the dry fly hook is relatively remedial, there was little to arouse anger. But hike up your waders! Out of nowhere the gloves come off with this blistering foreword: "Authorities darken counsel. An authority is a person engaged in the invidious business of stereotyping and disseminating information, frequently incorrect. An authority who lays down a law and dogmatizes is a narcotic, a soporific, a stupefier and opiate. The true function of an authority is to stimulate, not to paralyze, original thinking. But then, I suppose, he wouldn't be an authority." All of this with no mention of poor Halford, then seven years in his grave, who had never sought confrontation. I could only assume that the legions of Halford's followers must have placed a festering burr under Skues's saddle. Robson makes an interesting comment, suggesting that Skues's remarks, while bitter, reflected the spirit of the times following the Great War. "A distaste for authoritarian, pre-1914 systems and dogmas was evident in art forms, music and literature; it was the age of Lytton Stachey, of mocking eminent Victorians and of the new term, 'debunkingp" (*The Essential G. E .M. Skues*).

The reader, expecting a continuation of this diatribe, is instead invited to share in some deep and speculative thoughts. Much of the book is a compilation and melding of previous articles in *The Field*, *Fishing Gazette*, and the *Journal of the Fly-Fisher's Club*. Accordingly the content is a potpourri of well over a hundred variously related topics and sub-topics. The first part deals with the senses (taste, smell, vision, and so forth) of trout and their rise forms. (Ronalds began what became an almost compulsory obligation to reassess aspects of trout physiology.) The mid-portion of the book is a comprehensive treatment of fly dressings and the reactions of trout to various aspects of design as strongly influenced by the reflective and refractory qualities of the surface film. The remainder constitutes a miscellany of entertaining "fish stories" that causes the reader's focus to jump around a good bit in that there's no central theme.

The book is better than half finished before the chapter "Excommunication of the Wet Fly" comes along. Here barrister Skues constructs a "complaint" against the wet fly as in a civil case followed by an answer. "The practice of fishing wet flies in the Hampshire chalk streams has been shown to be an ineffective means of killing fish of decent size. Meanwhile anglers employing this method damage smaller fish, scare others making them shy and thrash the water to the detriment of the sport of others." Then with little emotion come arguments supporting each point in defense of the wet fly. Halford is treated fairly and with respect and it is shown that he considered the upstream wet fly inappropriate for the chalk stream only because he believed the method to be generally ineffective and never because it was inherently evil. Skues first attacks the major flaw in the complaint by agreeing that, fished downstream, as of old, only small fish are attracted to the wet fly, and many are "pricked" as well. He agrees that the method should be prohibited. However, fishing the wet fly upstream is an entirely different practice and one that, to Skues's knowledge, has never been regarded as offensive by any other angler. He goes on to show that the claim that wet-fly fishermen beat the water indiscriminately is not true of the upstream method and that the charge that wet-fly fishermen engage in hammering unresponsive trout is more characteristic of the dry-fly man. Skues recites a series of situations wherein the dry fly was ignored by the very same trout that rushed to take a wet fly. He goes on to quote the accepted fact that a mayfly spends nearly all its life in the larval form and is thus appropriately imitated by a wet fly, whereas the stage favoring the dry fly is evanescent. Skues rests his case logically: "There is no question of wet versus dry fly. Each in its place and used according to knowledge is surely the way of wisdom." This is followed by a recounting of three productive evenings on the Itchen. On one occasion the trout would only take the blue-winged olive dun, on another only the spinner and on the third, only the nymph. These few pages accurately reflect the book's content, for Skues pays as much attention to the floating fly as the submerged, just as in his first book. There is a final cranky remark concerning authorities: "It enrages me to think of all the good evenings' fishing I have missed through believing the pundits. The moral is, never believe a thing you are told about fishing until you have proved it, not only once, but over and over again." He refers to such false certainties as that there is no hope when there is an evening rise in the wind, mist, bright moonlight, or for some time after weed cutting, backing up each with an illustrative example.

THE HUMORIST

Halford wrote clearly, in considerable detail, and in a somewhat sterile, manner whereas Skues, a first rate raconteur, could easily have written for *Punch*. This was the era of the pseudonym, and Robson lists over a dozen of Skues's such as "Seaforth and Soforth," "Spent Naturalist," "A. Fluker," and his favorite, "Val Conson." (The title suggests that if an author submits something of value, it would be nice to be paid consonant with said value.) As mentioned, Halford sometimes wrote under the name "Detached Badger" and the pages of the *Field* and *Fishing Gazette* were riddled with many others. It's sad that the nom de plume has died, remembering Sparse Grey hackle (Alfred Miller) and his *Fishing Days, Angling Nights*.

Skues's book *Side-lines, Side-lights and Reflections* followed in 1932 and is largely a compilation of contributions to a wide variety of publications including *The Journal of the Angler's Club of New York*. Almost all his stories in the first books had been designed to establish or illustrate some point. They were often seasoned with a colorful description such as ". . . and in a moment a great fish leapt five and twenty yard away and fell back with a noise like a dog being thrown into the water." *Side-lines*, however, offers an array of paragraph-length pieces written with the sole intent of soliciting a chuckle. Some, such as the rather disturbing tale of Theodore Castwell, live on today. Briefly, Mr. Castwell, an avid fly fisher, then deceased, had been unduly aggressive, rudely spoiling the sport of others. When his time came, he audaciously defined a satisfactory situation for himself to assist St. Peter in coming to a decision about his future. Things got off to a heavenly start: a beautiful day along a classic chalk stream, large, rising trout everywhere, and a river keeper at hand with a box full of perfect imitations. The trout were consistently willing. At length Castwell says: "Well, I shall knock off now, I must have had about thirty brace from that corner." The keeper replies: "Beg pardon sir, but his Holiness would not like that." "What?" said Mr. Castwell, "mayn't I even stop at night?" "No night here, sir" said the keeper. "Then do you mean I have got to go on catching these damned two-and a half pound-ers at this corner for ever and ever?" The keeper nodded. "Hell!" said Castwell. "Yes," said the keeper.

Skues invented other preposterous characters to play the dupe in a series of stories. The Infernal Novice, conceived in 1912, appeared frequently. This friendly fellow was perfectly inept yet somehow caught large numbers of large fish with obscenities such as the Carrot Fly and Bread Fly. In "Cross-fertilization," the Novice searches for the best blend of the best characteristics of several patterns like a race horse breeder, e.g. "Crème de Menthe, by Corkscrew out of Green Bottle." Later Skues joins with the Novice in a successful effort to drive Castwell to distraction. Following him secretly, the Novice uses a BB gun to simulate rising fish and when a real rise comes on, peppers a pair of nearby swans into violence.

Skues targets a number of his many friends. Hugh Sheringham, an editor of the *Field*, had encouraged Skues to write both *Minor Tactics* and *The Way of a Trout* and better still, espoused the upstream wet-fly method. Nonetheless, Sheringham appears unflatteringly as "Scaringham" or "Piscator Rotundus." The gist is that in the process of stalking, anglers often hide behind tall grass or shrubs. Everyone knew when Scaringham had been on the stream for he left behind large "butts" or elephantine burrows of crushed vegetation.

Walter Coggeshall, an American member of the club, was much respected for his work in developing silk lines and worked with Skues in their design. However his preference for unusually large flies led to a wonderfully zany piece, "Surre fowls." The Member for America or

M.F.A. was somehow able to carry several of these large birds in his jacket, each attached to a "meat hook." When cast to rising trout, the Surrey fowl was viciously attacked, even by small fish. One account: "It rose to the danger like our men in France and Flanders, and dragged the fearsome fowl beneath the water. . . . " That fish managed to escape the meat hook but fell prey to the "clutch of these fearsome birds."

Coggeshall had an important influence upon Skues by introducing him to the Leonard rods with their precision-milled compound tapers. Skues's devotion to his nine-footer, the W.B.R. or "world's best rod," is said to have helped popularized the Leonards in England. He fished it to the end of his days. Skues believed that his wrists, injured years before in football, were weak, but his associates failed to detect any deficiencies in his casting.

Robson mentions an interesting historical note regarding the introduction of nylon monofilament on the chalk-stream scene. In 1938 and 1939 letters were exchanged between Skues and Baird Foster, an American interested in obtaining a marrow spoon. Foster had even built a bamboo rod modeled after the W.B.R. and sent Skues strands of the new material made by Dupont. Comparing it to gut, Skews had some reservations, but concluded: "I can see a good future for the material."

Nymph Fishing for Chalk Steam Trout

Skues's last work was published in 1939 when he was in his early eighties, ten years before his death. Having promised himself not to write another book, he reneged for several reasons. There had been a recrudescence of dry-fly purism in the 1930s. In the two decades since *Minor Tactics*, where the wet-fly nymph had been primarily criticized as an ineffective means of fishing the chalk streams, the focus of attack upon the nymph had turned full about. Now the nymph was seen as too deadly! The members of the Itchen Syndicate disapproved of Skues's successful application of the nymph (they were probably envious), resulting in Skues's agreement to refrain during the latter part of the 1936 season. His back was to the wall in 1938 during a formal debate at the very Fly Fishers Club he had served for so long. Terry Lawton's *Nymph Fishing a History of the Art and Practice* (2005) provides some details.

The evening began quite pleasantly with an introduction containing the kindest words for the nymph and for Skues. Indeed the tenor was such that there was no seeming need for a debate. Then came a vitriolic rebuttal by Sir Joseph Ball and various apostates who had once fished the nymph, most notably J. J. Mottram. [Mottram's stance against the nymph had become so strong that he had written the deprecatory *The Ethics of Nymph Fishing* a year before the debate.] Ball's arguments included the obvious—fish killed via a nymph are no longer at play for the dry fly man. Others were apocryphal—early in his career after thoroughly exploring nymph fishing Halford had rejected the practice. Another,—nymphing is a less delicate art and too easy since the problem of drag is much reduced. And some were simply ridiculous—"nymph-pricked" fish no longer take flies from the surface; even the sight of a nymph might have this awful consequence. And finally he testified that first-rate chalk stream fisheries had been ruined by the nymph.

Lawton said that Ball "revealed himself as a master of misinterpretation, misunderstanding and the half-truth."

The elderly Skues's self-defense in *Nymph Fishing* was clearly stated yet lacking in vigor. Aside from John Waller Hills, a great friend and believer who often fished with Skues and a few others, the debate was one-sided. The notion that Skues was actually cast out of chalk-stream society, a pariah, and denied access to certain waters is, however, incorrect. That there were snide remarks from those less successful is highly likely. The need to defend what was, after all, the reality, could only have left him with a certain bitterness, but his replies to his critics were dispassionate: "The dry-fly man, after the long innings he has had is often not of an open mind, and is unwilling to give up the exclusive doctrine of the dry fly, and he therefore seeks for arguments to justify it." As he had in *The Way of a Trout*, Skues attempts to summarize the complaints against the nymph. I've condensed his arguments into a list.

1. Nymph fishing is easier than the dry fly because drag is less defeating below the surface. The dry fly is therefore the more sporting.
2. The nymph is too deadly and leaves too few fish for the relatively finer art of the dry fly.

3. Persistence in the use of the nymph drives the fish to feed deep by killing off those which feed just under the surface. These are fish that would eventually have become prey for the dry-fly man.
4. Trout feed mainly on nymphs, but some of the more sporting and delicate fish eventually feed on the surface.
5. For some or all of these reasons the use of the nymph on chalk streams is ethically wrong.

Then comes a four-page rebuttal, addressing and defeating each claim in sequence. Skues repeatedly quotes Halford, refuting, clarifying, or putting a different spin on what he said. He criticizes "anglers who wish to re-establish the exclusive dominance of the dry fly" as

1. Men who do not understand and will not take the trouble to learn the art of fishing with a nymph;
2. Men who find the whole fascination of chalk-stream fishing in seeing their fly taken on the surface;
3. Men who suppose nymph fishing to be a "chuck and chance it" method;
4. Men who think it an unduly deadly method ("It is too difficult an art to be that, even if the charge were true.");
5. Finally (and these are the great majority), men who do not know what a nymph is and assume that the objects commonly sold by tackle dealers are reasonable imitations, whereas as most bear no resemblance to the real insect.

According to Robson, Skues believed that opposition to this form of fishing was primarily a reflection of an inability spot trout beneath the surface, a requirement for any competent nymph fisher. I haven't been able to find this directly stated by Skues, but it makes sense. Spotting requires time spent on the water and focused practice coupled with a great store of patience and a determination to avoid the easy seduction of fishing the water. It's easy to imagine that a dilettante might never acquire the skill, even on the pellucid, glassy chalk streams. (All the more remarkable, Robson attests that Skues underwent surgery as a youth, leaving a partial deficiency in his left eye.)

CHAPTER THIRTEEN

EPILOGUE

Halford and Skues came from remarkably similar backgrounds. Well educated, both enjoyed successful careers, Halford retiring comfortably from business at forty-five whereas Skues remained with the same law firm until 1940, a span of over sixty years. Both were purists, Skues in the ways he tied his nymphs and fished them. The difference: Skues reacted to criticisms or disagreements with interest and collegial enthusiasm. His relationship with Frank Sawyer, discussed below, is a case in point. Halford was devoted to his family and contributed generously to various worthy causes. Skues accepted responsibility for six younger siblings after the death of his parents and enjoyed many long-lasting friendships. Although never married, he is said to have dedicated a romantic poem to a certain, apparently unapproachable, young woman. (This is reproduced in Robson's biography.) Skues is said to have taken note of the well-turned ankle, but that's all we know. An argument could be made that excessive expression of the fishing gene, if such exists, interferes with the influences of the conjugal gene, if such exists. Late in his Itchen fishing Skues became disenchanted with new rules within the Abbot's Barton syndicate he had helped to found in 1922. There were issues regarding stocking, the number of guests a member might invite and a requirement that member and guest must share a rod. Accordingly he resigned, moving from his long-time home in the south of London to the tiny hamlet of Nader Vale, close to Winchester, where he fished the Nader River. Frank Sawyer visited him there in 1945. "To me he looked a little, fragile old man who wore a black skull cap to cover his extreme baldness and heavy lensed glasses to aid his sight. But though frail in body and aged in appearance I quickly found that his brain was very alert and this became more apparent as our discussion developed. His ability to sort out the corn from the chaff, so to speak, was brilliant, as also was his questioning." Sawyer, a beginner by comparison, remarked that Skues took careful notes all the while. Skues caught his last trout at age eighty-eight on the Wylie, a tributary of the Nader. "There was no hatch that day or rising trout. I thought I saw a dark body turn under the foam and raised the rod tip. . . ." His ashes were scattered along the Itchen. ◄══

Mottram, Renaissance Angler

*J*ames C. Mottram, a fellow member of the Fly Fishers Club, published *Fly-Fishing, Some New Arts and Mysteries* in 1915, hard on the heels of Halford's *Handbook* and Skues's *Minor Tactics*. Hills had credited Mottram cursorily for "much new and stimulating information" although offering no further illumination. Mottram seemed a bit part player until I discovered a chapter in Arnold Gingrich's *The Fishing in Print* which honored him as "one of the twelve peaks in angling progress" and on a level with the other greats. And Schwiebert agreed in his encyclopedic work, *Trout*, referring to Mottram as one of fly fishing's giants. Finding a copy might have been a problem, for curiously, according to Gingrich, the book wasn't especially popular. Luckily the Flyfisher's Classic Library came to the rescue with a fresh printing (1994). What little we know about Mottram's background is largely thanks to David Burnett's introduction. Mottram was born in 1879, educated in medicine with specialty training in radiology, and later participated in the early development of radiation therapy.

New Arts and Mysteries is an assemblage of pieces published earlier in *The Field* and several other leading periodicals under the pen name "Jim-Jam." Like Skues's books, it includes neither an index nor a bibliography. The table of contents identifies twenty-four concise chapters with eye-catching headings such as The Treatment of Mud, Paradise, The Colour Sense of Fish, Some Bird Friendships, and In AD 2014. The list promises an eclectic and entertaining read. And so it is.

2014?

I can't remember ever beginning a book at the end. This time I couldn't resist. What could Mottram be getting at? I'd read George Orwell's *1984*. It had a lot to do with controlling people and their thoughts. That sounded a little like Halford. Orwell was born in 1903, so Mottram could hardly have read his books. But sure enough, *2014* turns out to be an Orwellian lampoon targeting Halford. Maybe Orwell read Mottram.

2014 is an attack upon the excessively controlled, manicured, and pasteurized sport Mottram feared might evolve. Three gentlemen meet at the outflow of a sewage farm, now a prized club water: the manager, a fisheries specialist or super gillie, and a member who has come for a day of sport, who happens to be the chairman of the Association of Trouser Makers (recalling Halford's means of livelihood). The club supports an artificial mayfly hatchery much like a fish hatchery. The larvae have been programmed to emerge at a precise time so that a large number can be released to accommodate a member who plans to be on the water, for example, "between 11:15 and 12:45." The pampered angler then fishes actual insects that have been candied or coated and glued onto the gut leader a short distance above a treble-hook or "hooker." Each angler pursues a particular highly educated trout that has been given a name, e.g., "Brown Beauty." Due to the trout's extreme shyness the finest gut, 9X or 10X, is required. The fly is floated downstream over a precise distance measured in inches, while an ingenious Y-shaped device attached to the mummified fly causes it to reach the fish before the "hooker." If all goes well the gillie rushes to get below the fish, shouting, waving his arms and throwing rocks or "stoning," thus chasing the quarry back upstream for easier capture. A trout can only be killed after it has been caught three times by the same member. If released, a tiny silver tag is pinned onto the fish, recording the angler's name, date and whether this was victory number one or number two. At the time of the fatal third event the fish is mounted for display with all its tags in place. Brown Beauty weighed almost eight pounds and was caught nine times by six members, the last of course by the Trouser Maker. It's pretty harsh. Pondering Mottram's relative lack of popularity, Gingrich suggests that he may have come across as unduly iconoclastic at a time when Halford was so much revered.

SUPPORTING SKUES

Even before *Minor Tactics* appeared, Mottram had been influenced by Skues. The very first words in the chapter "Nymphs and Bulgers" give us a clue: "I believe the day is not far distant when the dry-fly fisherman may have to give up or share his throne with another, and to acknowledge that it is as difficult to deceive a trout in its own element as in the air, if not more. I am sorry to say the dry fly hit the wet when it was down. I hope the wet has forgotten that." Mottram is referring to the old, downstream, multiple-fly method as "down" and goes on to identify the "modern wet-fly fisherman" as one who casts a single fly to a single fish feeding underwater, the fly a representation of the insect the trout is presumably taking. As to making the trout shy, Mottram suggests that in waters heavily fished with the modern, upstream method, the trout might rise for a floating fly more readily than ever. (I assume because they would have become aware of the dangers of the nymph.) In any case Mottram especially enjoyed the sport of pursuing shy fish. For him they couldn't be too fearful. In reality he preferred the dry fly, yet was always alert, willing, and able to play the other game. Fishing a large New Zealand River in the early 1900s he was told that there was no chance despite a heavy hatch of mayflies and actively feeding trout. That first evening the locals proved correct. Mottram couldn't get a single rise to any of his dry-fly patterns, only later realizing that the trout had also refused the drifting duns. "The next evening I went fully armed with nymphs and slew many fish." Terry Lawton suggests that Mottram was likely the first to introduce the nymph in New Zealand (*Nymph Fishing: A History of the Art and Practice*).

SKEWERING SKUES

It wasn't all arms around the shoulders. Mottram and Skues had some fundamental differences as displayed in articles that criss-crossed in contemporary journals. They did agree that "floating nymphs" were virtually inert and taken with a "dimpling rise" (Mottram's term), that bulging trout were sometimes present in numbers when relatively few duns were afloat, and that the bulgers were after larvae warming up for emergence. However Mottram suggested that the pre-emergent larvae were themselves active swimmers and hence the nymph too should be swum. To prove the point Mottram liked to fish to bulgers across,

across and down, or occasionally downstream and often took his fish on a dragging fly. Of course Skues strongly disapproved. Mottram's simple swimming nymphs were devoid of hackle since the legs of the naturals were folded back tight to the body and featured an expanded thorax profile. Skues felt that the delicate fibers, breathing in the current, added a lifelike quality.

The most significant distinction between the adversaries was Skues's preoccupation with mayfly larvae/emergers drifted topside in contrast with Mottram fishing his flies deeper and with more action and seemingly being less wedded to the mayfly. For example Lawton offers an illustration of four of Mottram's nymphs, all stressing silhouette. The Resting nymph's delicate gill structure stands out as compared with the Swimming nymph's streamlined taper. The Caddis nymph is similar, but with backswept antennae and beard hackle while the Diptera nymph's unique outline is quite stark. Admitting that he had never tried them, Skues saw "no special advantage in the silhouette fly patterns advocated by that very interesting writer and skillful angler, Dr. J. J. Mottram." In theory his nymphs seem to me to have the defects of rigidity, density, and dullness—and moreover are used dragging. Elsewhere he even described the silhouette nymphs as "dead as mutton" (*The Way of a Trout with a Nymph*). While he is nowhere near as dogmatic as Halford, we see here a certain prissy rigidity in Skues. These skirmishes with Mottram are mentioned rather frequently in *The Way of a Trout with a Fly*. It's very likely that Mottram's provocations such as the emphasis upon silhouette were part of the stimulus that led to Skues' second book. (Lawton found Mottram's nymphs the more realistic and discusses the feud more thoroughly in *Nymph Fishing*.)

The chapter *Fishing on Sight* begins with a particularly sarcastic slap at Halford and a peck at Skues: "This title will no doubt give a moral shock to the dry-fly purist. If he would protect his nice nerves from further shock, let him be advised to read no more. The fish stalker must learn always to focus the bottom of the river; he must search every foot, forgetting for a time that he is fishing, remembering only that he is hunting." None of our authors to this point had mentioned streambed trout in a hopeful way, quite the opposite, yet Mottram suggests that any fish of acceptable size should be pursued regardless of depth and whether or not it is actively feeding or in position. In pursuing this theory and practice Mottram anticipated two major figures in the evolution of the wet fly/nymph, his country-man Frank Sawyer and the American James Leisenring. Curiously though, while Mottram recognized that trout lying on a streambed often took drifting nymphs, instead of dropping his fly far enough upstream for it to drift to the fish, he'd try to get them to come up to a "Rabbit," an ant-like attractor tied with black fur and scarlet wool. (His flies and leader were not weighted.)

EXPERIMENTATION, REASON AND PRACTICAL APPLICATIONS

Mottram's strongest claim to fame was his prescience and creativity in pattern design based upon observation, hypothesis, and experimentation. "Fishing," he wrote, "is a great series of experiments." Among other innovations, the incorporation of soft feathers, such as marabou, to provide living movement in imitations of minnows and also mayfly larvae (a trailing shuck) is credited to Mottram.

He assaulted the murky challenge of whether trout react to color as humans do through a comparative anatomical dissection and microscopic study of the human compared with the trout eye. He reasoned that if the specialized cells of the retina are similar in type and arrangement, there should be strong similarities. Anatomically, that proved to be the case, but since that *should* was only a hypothesis, it needed testing. Accordingly, Mottram trained an aquarium fish to distinguish, after wetting the fly, between white, blue, and red, even in different shades. Mottram nonetheless considered color to be important only in the case of a floating fly when reflected light from a low angle was revealing or with regard to the color of the translucent portion of a floating fly. In his chapter *Some Optical Problems*, Mottram extended Ronalds's observations. Returning to the trout's "window," he employed principles of reflection and refraction angle as influenced by trout to angler distance and depth v. height variables. He described a "dark ring" formed by prismatic colors at the window's edge and most importantly, explained why a trout's first image of an approaching mayfly dun must be of its prominent "sailboat" wings, at first just the tips.

Acknowledging that Halford first stressed exaggerated wings, Mottram's scientific observation is thought to have become the basis for the "tall wing" patterns popularized by Vincent Marinaro in his *A Modern Dry-Fly Code* thirty-five years later. The extent to which Marinaro credits Mottram is unclear; he merely states that Mottram's book "suggested some interesting applications of new principles in the art of fly dressing." As we know, dependence upon wing profile was perpetuated in Swisher and Richards's hackle-less dry flies and the now standard "Compara-dun" pattern style (*Selective Trout*, 1971).

In his chapter "Flies of the Future" Mottram singled out the features that characterize an effective dry fly and listed them in order of importance. He considered form or silhouette closely resembling that of the natural insect most essential, necessarily followed by buoyancy and "lightness." In designing an imitation of a mayfly spinner with a clear abdomen, he simply left the center of the shank bare. Mottram's focus upon silhouette was so great that he designed wingless patterns featuring realistic body shape, legs, tails, and even gills. The same theme was carried over to nymphs (the concept of the "floating or resting nymph") and wet flies, the latter to include ants, beetles, and minnows.

SMUTTING

Mottram's precocious imitations of midge pupae were tied with a tiny ball of cork that hung in the film. (A bit of foam enclosed in a delicate net serves the same purpose today.) Another pattern innovation featured a tail of gray turkey down, representing a trailing shuck. Unlike Halford, Mottram rather welcomed the challenge of the smut. "When dealing with smutting fish, the angler should have many strings to his bow." He suggested a very simple wet fly with a bare shank and a bead of black thread and a collar of short hackle behind the eye. This was cast into the film much like a dry fly. Although much larger than the naturals, it was predictably effective, just as we sometimes use outsize flies to suggest midge clusters.

The insightful Mottram's autopsies also showed that trout, supposedly smutting, could well be taking tiny terrestrials such as aphids, leaf rollers, or the larvae of the saw fly, dropping from leafs overhanging the water. Once again, Marinaro discovered the same phenomenon in Pennsylvania chalk streams many years later (*A Modern Dry Fly Code*, 1950).

A Man for All Seasons and Hemispheres

As in the case of Ronalds it seems justified to afford Mottram the title Renaissance Angler, given that in one chapter or another he assumes so many identities. He was surely an adventurer. Mottram's experiences in New Zealand (Tasmania too) are a prominent part of the book and so enthralling that you'd expect it to have been a best seller. This was in 1909 when "good fishing throughout the islands was common, accommodations were not so." He brought back wonderful memories of my visits to rivers such as the snow-fed Clinton on the South Island. Mottram had a tough day there, only four trout (on dry flies), and then having to lug over twenty pounds of fish flesh (dressed) back to camp! Catches like this presumably helped defray Mottram's expenses. I couldn't match his stories from the Clinton, but we suffered equally from the long-lasting, long-itching sandfly bites that are an indelible part of Kiwi fishing.

Lake Taupo on the North Island was a hot spot then as now. One angler is said to have killed nine tons of trout in a single season at circa ten pounds each, roughly 1,800 fish. Something of a fisheries biologist, Mottram came back with a strong preference for the brown over the rainbow. He feared that the rainbow had a proclivity to spawn so bountifully that the fish would become small and unfit and suggested measures to discourage spawning in streams feeding into the big lakes. And while the browns fed contentedly on mayfly duns, as they should, rainbow were too easily distracted by passing minnows or crayfish. Mottram also noted that "rainbow fight with their tails, brown trout with their heads."

He qualified as an anatomist as well. In addition to his studies of the trout eye, a chapter is devoted to autopsies. The illustrations and concept were surely borrowed from Halford, but the instructions are far more detailed and the anatomical parts correctly labeled. Mottram's knowledge of the botany applicable to fishing was amazing. He classified "water weeds" into three categories, beneficial, harmful, and intermediate, on the basis of whether a particular plant provided a food source for the creatures (emphasis on mayfly larvae) of interest to the fly fisherman. "Bad" plants got that rap by crowding out more desirable species, covering clean gravel or feeding beds, or snarling and tangling lines. It followed that the undesirables should be uprooted when possible, meanwhile encouraging others such as the River Crowfoot (*Rannunculus fluitans*). Yes, he matched common names to the corresponding genus and species.

Mottram was an expert ornithologist, sometimes tying feather to fin in his stories. He wrote of a pleasing experience wherein his imitation of a mayfly dun was accepted at the same time by a large trout and a swallow. The first cast dropped the fly upstream from the fish. The bird quickly picked it off the water, flew off, and dropped it. Second cast, same bird, same outcome. On the next presentation, fly, trout, and bird collided, and the big brown withdrew in consternation.

Describing Mottram as a physicist might seem a stretch, but isn't the study of radiation energy and its application a branch of physics? His contributions pertaining to the "window" are also consistent while he applied a simpler sort of physics to rod design wherein angles, weights, and vector-of-force calculations showed that in playing a fish, a shorter, whippier rod is more powerful than a longer, more rigid one. *Naturalist* might be the best overall title for Mottram. Whenever the fishing turned sour, it seems he equally enjoyed the lessons and mysteries provided by the streamside biota. His words show that he often thought of Darwin, especially when he was in New Zealand.

Some chapters treat of matters purely practical. I suspect that a lot more of us have thought about carrying a tying kit on the water than actually carry through. For Mottram, though, this was standard practice. He used a watchmaker's thumb vise, a hackle pliers, and a small store of materials. Ready-made innovations solved all sorts of on-stream mysteries and frustrations when trout were taking an unanticipated insect. Simple impressionistic patterns usually did the job. He believed that in playing a fish, control of the line and reel requires more finesse and touch than managing the rod and that the right hand is the more capable for most of us. Hence Mottram suggested that the reel should be mounted with the handle on the angler's right as many have come to prefer. On bowing to a jumping trout: "The jumping fish can almost always be stopped by holding the rod tip out laterally and close down to the surface—it acts like a charm." Other chapters devoted to casting, stalking, gut leaders, and dry fly fishing lakes contain information that was not necessarily new at the time.

Should Mottram be inducted into Fly Fishing's Hall of Fame? His general lack of recognition could support a thumbs-down vote, and unfortunately his legacy rests almost entirely upon one book although he wrote others later. In favor, like Ronalds he actively pursued an array of remarkably eclectic interests. Backing up Gingrich and Schwiebert, Lawton further attests that we can trace a number of the roots of modern fly-fishing practice back to Mottram, especially aspects of the design and presentation of nymphs (*Nymph Fishing in Theory and Practice*).

THADDEUS NORRIS AND THE
EARLY YANKEE FLY FISHERS

CHAPTER 15

T he cornerstone of American angling literature." Arnold
Gingrich awarded that honor to *American Angler's Book*
published by Thaddeus P. Norris in 1864 (*The Fishing in
Print*). John McDonald's summary was more descriptive:
"The great American fisher of the century was 'Uncle' Thad Norris,
learned in fishing literature and experienced in native practice. He
knew about everything there was to know in his time, put it all down
in 1864 and thereby equipped the school of early American fly-fishing
with a rounded theory and practice" (*Quill Gordon*). Norris was well
educated, an outdoorsman and highly successful businessman engaged
in the mercantile trade in Philadelphia. His writing style was delight-
fully unique. Several earlier authors nonetheless helped set the cornerstone in place, and to the extent that they influenced Norris, we
should understand their stories.

THE REVEREND BETHUNE

It's pretty much agreed that British officers and governmental officials introduced fly-fishing to the colonies. Paul Schullery tells us
that the artificial fly was successfully employed at the time of the
revolution near larger population centers, and yet it's difficult to find
much in the angling literature from the early postcolonial years. In
1847 George Washington Bethune edited the first American edition

of *The Compleat Angler*. (He was a minister of the Dutch Reform Church, and fearing that the congregation might be offended by his involvement with a frivolous entertainment, he identified himself modestly as "American editor.") In great part Bethune's importance lies in his *Bibliographical Preface and Notes*, as they describe fishing American waters at the time. According to Schwiebert, Bethune was one of the first notables to catch native (brook) trout on flies. He was fond of the Brodhead, a small river arising in the Pocono Mountains of southeastern Pennsylvania. Running through the village of Analomink, the stream links historical events to our sport. Captain Daniel Brodhead obtained acreages in the valley in 1734 where he and his six sons built a combined fort and manor. Displaced Delaware tribes swept through the area during the Indian wars of 1755; the Brodheads stood fast. Each of the boys eventually served as an officer under Washington. The Brodhead (alternatively, *Brodheads*, without an apostrophe) was also an exceptionally fertile fishery, later attracting leading anglers from New York and Pennsylvania such as Bethune and Norris (*Trout*, 1978).

In the Yankee spirit, Bethune found no need to blend or compare his experiences with what various English authors might have had to say. He took up the issue of imitation versus nonimitation at some length, favoring the latter. "The presence of the hook per se defeats any pretense at true imitation." He also descried the traditional English practice of identifying the best patterns for each month and suggested that a relatively small number of flies would suffice as "attractors." "When drawn along the water," he wrote, the fly "has the appearance of being a living insect, whose species is quite unimportant, as all insects are equally welcome." While he mentions some familiar flies such as the Cowdung, Bethune recommends fifteen patterns, with recipes and commentaries listed by number rather than by a name. Some were "tail" flies, others "droppers," and all, as far as one can tell, were "made in America." Other bits of advice are still valid such as "have *nothing shining* about the rod, as the flashing of light will certainly scare the trout." Contemporaries, Bethune and the Norris were friends and likely shared experiences on the stream. The new edition of *The Compleat Angler* brought the pious Walton and less pious Cotton to the attention of a much wider American readership. Walton, much admired by Norris, appears frequently in the *American Angler's Book*.

John Jay Brown's Weighty Compendium

A New York City tackle dealer, J. J. Brown took on the huge task of summarizing the fishing of the same period in *The American Angler's Guide or Complete Fisher's Manual for the United States*, published in 1845. This is a collection of pieces by "experienced anglers of both hemispheres," over twenty in all. Trout Unlimited commissioned Derrydale Press to reprint the work under the title *Fly Fisherman's Gold* in 1993. With something over 300 pages, the work is truly encyclopedic. A more accurate title might be something like *An Angler's Practical Ichthyology*, for after a brief discussion of biology, baits, and tackle, some fifty fish species are discussed. Pike, bass, carp, catfish, chub, dace, perch, roach, tench plus a number of saltwater fish, and just about every possible means of growing, catching (from artificial flies to ground bait), and cooking them are covered Attractive line drawings support the descriptions of each fish. It makes you wonder if, whether for food, sport, or more casual entertainment, anglers in those simpler days didn't pay more attention to "coarse" fish such as the modest chub and bream. Only one chapter (if a long one) is devoted to trout. Salmon are the topic of three others. In point of volume, there is relatively little gold for the trout fisherman, and unfortunately, while Brown accumulated the comments and opinions of a scattering of American anglers, he quotes Walton, Cotton, and the "North Country Angler" so frequently and at such length that it's difficult to winnow out what might have been fresh, indigenous contributions. Regarding fly fishing Brown commented: "From the fact of there being comparatively few who practice with the fly, some English writers are of the opinion that there are no fly-fishers in America, and many of our own countrymen think there are very few; but this is a great mistake. There are hundreds of good fly anglers, and many that can throw a fly with the most experienced in Europe. In some parts of our trouting districts there are many ladies that can throw the fly with as much dexterity and grace as those that are made of sterner stuff." He meant that last sentence as a compliment, but today that sentence would be edited.

Brown's patterns show a strong English influence. About half carry the same names as Ronalds's list (Cow Dung, March Brown, Green Drake, Cinnamon Fly), and the recipes, if not identical, are very similar. His descriptions of various trout species necessarily cover brookies. There was the "common" as well as the "silver" trout, the latter a strain that became extinct. The ponds and briskly flowing streams on Long Island provided the finest examples, weighing four pounds and more! The "black trout," usually found in muddy, sluggish streams or large ponds with clay bottoms in the "roughest and wildest parts of our country" (western Pennsylvania and some parts of New York, Vermont, and New Hampshire) were regarded as inferior game on the rod or table. Brown attested that large "sea trout" fed in the brackish waters of tidal rivers. (According to Behnke, brookies indeed migrated into bays and estuaries from the Atlantic States on northward [*Trout and Salmon of North America*].)

Among many pieces of advice for the frontiersman angler I especially like this one: "When there are trees or bushes very close, I advise the bush angler to take a hedge-bill or hatchet, and cut off two or three branches here and there, at proper places and distances and so make little convenient openings, at which he may easily put in his rod and line; but this is to be done some time before you come there to fish." I recently waded several tree-lined streams in Ireland with a guide who carried a pair of heavy-duty clippers on his belt. He selectively lopped off overhanging branches, even small limbs. On several occasions my fly was recovered in this way.

The *American Angler's Book* offers an extensive ichthyology very similar to that in *Fly Fisherman's Gold* through a series of chapters featuring "families" such as Pike, Perch, Carp, Herring, and Salmon. Fortunately, however, Norris allocated the lion's share of attention to trout and salmon.

Norris's Cornerstone

The *American Angler's Book* is a weighty work of just over 600 pages with a price tag of about $150 for a first edition in fair condition. Always direct and quick to the point, Norris gives the reader a clear concept of what is to follow in his introduction.

> With a view of filling up the blank left by my predecessors, of correcting some erroneous ideas that have been imparted, not only concerning fish, but the adaptations of English rules and theories,

without qualification to our waters; and with the object of making the angler self-reliant, and to encourage him as much as possible to make the best of such resources as may be within his reach, especially as regards his tackle, I have devoted many spare hours to the following pages; in writing which, to use the words of Issac [sic] Walton, "I have made a recreation of a recreation."

Glen Law's description of Norris hits the mark: "The True Angler, Norris's Archetype, is an image of the self-reliant, innovative, capable, sure American, the same who forged into the frontier, founded a new nation from nothing and killed him a b'ar when he was only three" (*A Condensed History of Fly Fishing*). Norris's words ring of fresh beginnings and departures that shed the trappings and traditions of our oppressors, not long removed. A bit further into the introduction Norris reveals still another side: "I have at times indulged my sense of the ludicrous or the ridiculous." He refers to passages written in a conversational manner, homespun and sometimes seasoned with ironic humor something like the early Mark Twain. The very first chapter sets that tone under the subtitle "Different kinds of anglers": "Anglers may be divided into almost as many genera and species as the fish that they catch, and engage in the sport from as many impulses." We meet nine fishermen variously dubbed Fussy, Snob, Greedy, Pushing, Spick and Span, Rough and Ready, Literary, Pretentious, and finally, the True Angler. The invitation to think about how you and your friends might fit here or there is quite compelling. The Spick-and-Span Angler has a highly varnished rod and a superabundance of useless tackle, whereas "The Literary Angler who reads Walton and admires him hugely; he has been inoculated with the *sentiment* only; the five mile walk up the creek, where it had not been fished much, is very fatigueing to him; he did not know 'he must wade the stream,' and does not until he slips in. . . ." By contrast,

> the True Angler is thoroughly imbued with the spirit of gentle, old Izaak. He has no affectation, and when a fly-cast is not to be had, can find amusement in catching Sunfish or Roach and does not despise the sport of any humbler brother of the angle. A true angler is generally a modest man; unobtrusively communicative when he can impart a new idea. . . . With many persons fishing is a mere recreation, a pleasant way of killing time. To the true angler,

however, the sensation it produces is a deep unspoken joy born of longing for that which is quiet and peaceful, and fostered by an inbred love of communing with nature, as he walks through grassy meads, or listens to the music of the mountain stream.

Norris rises to the defense of the chub and other widely disparaged "rough fish." He says of catching bream: "With a pleasant companion, a bottle of claret, ice and cold fowl, the day passes pleasantly enough." That's Waltonian. Norris also anticipates the future with a fine (illustrated) treatment of Redfish in the Gulf of Mexico, an adversary currently of great interest to a number of my angler friends. (The text is sprinkled with occasional illustrations, some small and decorative, while others take up a full page and are quite detailed.)

About Trout

Three chapters are given over to trout: appropriate tackle, "fly making," and stream tactics, some seventy pages in all. A twelve-foot rod weighing at least twelve ounces with a stiff tip is favored. "In giving a list of flies best adapted to American waters, I have done so without reference to the opinions of English writers, considering many of their rules and theories regarding flies inapplicable to our country." Instead Norris turns to his own experiences and those of his friends, concluding, "an extensive knowledge of flies and their names can hardly be of much practical advantage."

Here Uncle Thad ponders the choice of a fly pattern: "What a pretty fly—half sad, half gay in its attire, like an interesting young widow, when she decides on shedding her weeds, and 'begins' to take notice." Then for his dropper he turns to a more somber fly: "What a plain homely look it has; it reminds me of 'the Girl with the Calico Dress.' You are not as showy, my dear miss, as the charming young widow, but certain individuals of my acquaintance are quite conscious of your worth." (That's a little risque compared to Walton.) Elsewhere he comments on the attractive curve a rod develops when playing a good trout and likens it to the curve one might admire in "the outline of a woman's drapery." Sounds like Cotton. (Lest we get the wrong impression, Norris lived happily with his wife and three children in Bryn Mawr.) The old "fly of the month club" dating back to Dame Juliana really got Thad's goat, as it had Bethune's. Norris refers to Bethune as "The Divine" and puts him to use as an example of an angler as opposed to a fisherman. "The angler uses the finest tackle, and catches his fish scientifically with an artificial fly, and he is mostly a quiet well-behaved gentleman. A fisherman, Sir, uses any kind of 'ooks and lines and catches them any way, so he gets them, its all one to 'im. And he is generally a noisy fellah, Sir, something like a gunner."

Norris favors "hackle" patterns over winged flies, although not to the exclusion of certain English staples going back to Ronalds, e.g. the Grannom, March Brown, and Green Drake. The chapter on "Fly Making" begins apologetically by suggesting that the student would best learn hands-on from a professional. The twenty pages nonetheless offer sound instruction and cover every aspect. Norris concludes with a note of encouragement: "Do not throw away all your first attempts for a much rougher-looking fly than you suppose will kill." The reader is encouraged to explore fur and feathers wherever he finds them, even clipped from his wife's wrap, and not to pine for exotic materials. Uncle Thad said about the tyer's "gentle craft": "There is no in-door occupation so absorbing and time-killing, and one forgets in it little annoyances or heavier cares, and almost finds at home a substitute for the pleasures of the stream." It is a bit surprising when Norris refers to the "abominable permanent vise, screwed to a table and reserved for the angler with ten thumbs upon his hands." Uncle Thad preferred to tie with a hand-held "pin vise"; if one's fingers are long and pliant, the knack can be acquired with perseverance. (Like Pulman, he gives the Double Angler's or Surgeon's Knot high praise for leader splicing with no mention of the Blood Knot.)

TACTICS

Chapter XII, "Trout Fly Fishing—the Stream" contains the instructional core for the trout fisherman. (Salmon are discussed in other chapters.) The beginner is reminded that "it is not strength, but an easy sleight and the spring of the rod that effects [sic] the long and light cast—light as falls the flaky snow." Norris repeatedly mentions the attraction to the brook trout of a shaded bank and the wisdom of a horizontal cast, "back hand" when necessary. He was very much a presentation-over-pattern man, meaning action imparted by a dropper constantly teasing, skimming, jumping. "The stretcher fly which follows in its wake may be allowed to take care of itself, for, as a general thing, it matters little whether it is on or beneath the surface." As that suggests, Norris was a top water fisherman, and he generally liked a quartering, downstream cast, admitting that the dropper also "plays well" directly below. Wading against the punishing currents of swift streams was impractical, "I have seen beginners entangle them (the flies drifting back) in the legs of his pantaloons." Some instructions invoke a visual image: "When a large rock projects above the surface in water of sufficient depth, the angler should cast near the edges on both sides, then above where it repels the force of the stream; or he may have a rise in the eddy just below, where the divided current unites again." "A deep bend in the stream where a caving bank overhangs, affords a likely cast, especially where stumps, logs or drift wood lie about. It is not a bad plan when fish have risen and refused one's flies in such a pool to sit patiently down and change them for smaller ones." With these and a dozen other prudent observations the beginner can pick up enough practical instruction to wade in with a game plan.

Uncle Thad and the Dry Fly

Was he the first in America? Fishing downstream wouldn't preclude attempting to keep the flies on the surface, even classic wet-fly types, and after all, every attempt was made to tease with the dancing dropper fly. But that goes back to Cotton. However Norris did write: "If it could be accomplished, the great desideratum would be to keep the line wet and the flies dry. I have seen anglers succeed so well in their efforts to do this by the means just mentioned, and by whipping the moisture from their flies, that the stretcher and dropper would fall so lightly, and remain so long on the surface, that a fish would rise and deliberately take the fly before it sank." (The "means just mentioned" seems be a short cast with a long rod and very light line.) According to Schwiebert, Norris fished the Willowemoc with the express intention of keeping a brace of flies on the surface and caught fish *Trout* (1978). But in Norris's book it was a friend of Uncle Thad's who accomplished the feat. They were fishing a pool below a dam on the "Williwemock."

The fish were shy and refused every fly I offered them, when my friend put on a Grannom for a stretcher and a minute Jenny Spinner for a dropper. His leader was of the finest gut and his flies fresh, and by cracking the moisture from them between each throw, he would lay them so lightly on the glassy surface, that a brace of Trout would take them at almost every cast, and before they sank or were drawn away. He had tied these flies and made his "whip" especially for his evening cast on this pool.

Thaddeus, always a good sport, assisted by netting and "in a half hour counted several dozen." We don't know who this friend may have been, but it was he, not Norris who with specific intent cast to rising fish with flies designed to float and false cast to assist toward that end. Norris does not indicate that he necessarily adopted the practice himself. John McDonald preferred to regard Norris' experiments merely as "a gesture" (*Quill Gordon*). I had the same feeling. He found his friend's success more an interesting novelty than an innovation.

Difficulties occasioned by the impending Civil War, "the present unhappy rebellion," delayed the appearance of Norris's book for some years. Thus it could be argued that the dry fly appeared in America at about the same time as Pulman described it in England. Theodore Gordon, the great Catskill fisherman, often thought of as the father of the dry fly in our country, called Norris's work his "book of books," the primer which started him on his career. (An expanded edition of the *American Angler's Book* was released after the war.)

A Philosophy Too

The *American Angler's Book* constitutes a fisheries biology, broadly defined, with special attention to the catching of salmon and trout. But it's far more than that. When you finish the book (even skipping over some of the ichthyology), Norris seems as much a philosopher and insightful student of his fellow man as an angler. The "Houseless Anglers" were a group of about ten of Uncle Thad's angler buddies. The members had been gradually assembled through chance meetings of kindred spirits on the stream and mutual agreements to fish together again. Through interconnections these individual friendships evolved into a loosely knit club of sorts, but where to meet? On the water, of course, and for what better purpose than the "noonday roast"? Lunch would be taken care of with ample time for fellowship at an hour when the trout were least likely to be active. (By pre-arrangement, some members fished down to the dinner site and others, up.) A loaf of bread, hollowed out, contained butter with room for salt and pepper and matches in a pocket above the high-water mark. The fresh-caught trout were either fried on hot stones or baked, directions given by Uncle Thad. Rested and refreshed by a good pipe or two, the Houseless Anglers enjoyed even better sport during the afternoons. You never know what's coming from Norris. He was very fond of the soothing, mind-clearing benefits of his pipe and "seegars" and wrote, "What a pity infants aren't taught to smoke."

Finishing the book's last chapter, the reader is surprised to find that about a hundred pages remain in a section entitled "Dies Piscatoriae." It's an appendix of a kind. Norris didn't like that word, though, finding it to be like the tail on a dog, "useless" and "an excuse to close a book's covers." "I therefore discard that stale old word, 'appendix' and use the new bait . . . to lure the reader on to the end of the book." It turns out that the "Dies" records the conversations and events of four "Noonings," that is, the deliberations of the Houseless Anglers. The back and forth dialogues remind one of Walton or Cotton, although in the Yankee vernacular. "Fishing Alone" is another philosophical segment where Norris, a pious man, concludes with thoughts about an angler's Sabbath, quoting Walton. Combining his own adventures with those of the Houseless Anglers, Uncle Thad's book also served as a pioneering guide to wilderness areas from the streams of the rugged Adirondacks, then onto Pennsylvania and even Michigan's as-yet-unspoiled Upper Peninsula with its grayling fisheries.

The Brookie's Demise

For the Houseless anglers the 1870s began a distressing time. Brook trout were disappearing from their natal waters at an alarming pace. Accounts of catches recorded at fifty and more pounds in a day, with individual brookies up to four pounds, had come from wilderness waters not so many years before. Pessimistic editorials suggested that trout fishing, as a sport, would be no more, and fishing clubs began to disband. Ernest Schwiebert writes:

The impressive forty fish baskets of large speckled trout had taken their toll, along with the rapacious lumbering that ravaged the magnificent virgin pine forests to provide railroad ties and framing lumber and mine timbers. It was a carnal waste that felled the hemlocks, stripping their bitter bark for the acid required in the tanneries, their naked trunks left rotting in the woods. The forests were culled for their hemlocks to process the flood of buffalo skins and cowhides coming east form frontier railroad towns like Abilene, Wichita and Dodge City. The uncontrolled lumbering and the clearing of the land for orchards and farms had changed the eastern trout streams forever. (*Trout*, 1978)

Norris remarked, "It is an old story everywhere along our mountain streams, of how abundant trout *once were*; and the angler is shocked and disgusted on every visit, with the unfair modes practiced by the natives and pot fishers in exterminating them." His was the last generation to enjoy bountiful "square tail" fishing. Behnke indicates that ova of trout, salmon, and char were transported to New York City in 1864. The ova that hatched were then unsuccessfully transported to streams on Long Island (*About Trout*, 2007). It was another twenty years before Browns were firmly established in this country (*Trout and Salmon*, 2002).

OTHER ACCOMPLISHMENTS

Norris's writings, while mainly concerned with angling topics, included a piece entitled "Negro Superstitions," which told the story of Br'er Rabbit and the Tar Baby and may have been the source for Uncle Remus, later used by Joel Chandler Harris.

Norris had a strong interest in all aspects of rod design, including a chapter on rod making in his book. It was only natural that he met Samuel Phillippe, a fellow Pennsylvanian and rod maker. Phillippe is generally credited with developing the four- and six-strip bamboo rods that Leonard later perfected. Norris must have learned from Phillippe, as they became friends, fishing the Brodheads together. For many years Norris built rods largely for his own fishing, preferring malacca cane (from a palm). However, sometime after 1860 he converted to bamboo and opened a small business. Norris Rods had apparently become popular by the time of his death in 1876.

ALSO WORTHY OF MENTION

Robert Barnwell Roosevelt, apart from being Teddy's uncle, was a complex and colorful fly fisherman of the same era. Paul Schullery introduces us to Roosevelt as a congressman, Minister to the Netherlands, reformer, novelist, New York State Fish Commissioner, naturalist, eccentric, and one of New York's most gossiped-about social lions. Roosevelt became a rival of sorts for Norris largely through his book *Game Fish of the North* (1862). In brief, he had some understanding of the entomology of fly fishing and correlative imitation. He knew Ronalds's work well, revising some of the English patterns that could be tied with materials available locally. Schullery noted that while Roosevelt's entomology was not systematic, it did include the major fly types and dates of emergence. Theodore Gordon later gave the work credit and praised Roosevelt's casting skills. The notion of bending over the shank of a Limerick hook to form a loop eye, the predecessor to the eyed hook, is also credited to Roosevelt. Reminiscent of the debate between Stewart and Francis over the merits of Borders and chalk-stream trout, Roosevelt touted the sporting qualities of "cultivated" Long Island fisheries as compared to those of the "rarely visited forest" where Norris fished (*American Fly Fishing*).

Frank Forester was a Cambridge-educated Brit who came to this country as a sporting writer. He developed quite a following during the years preceding the Civil War through his books and pieces in the American Turf Register. Nonetheless, the quality and accuracy of his writing, in the manner of the "dime novel," was not of the highest order. Gingrich suggests that this is why almost none of his work survives, although he was the most popular fishing writer for a time and one of the first to call attention to saltwater opportunities for the fly fisherman. ━━►

THEODORE GORDON AND THE TRANS-ATLANTIC DRY FLY

CHAPTER 16

*E*astern fly-fishermen of a certain maturity properly view the Catskill streams as sacred heritage waters and Theodore Gordon as the saint in residence. Agreeing that Gordon helped introduce the dry fly to the Neversink, Beaverkill, and Willowemoc, Paul Schullery notes that he "has been glorified more with each passing generation, until within the past ten years, he has been given credit for, among others, the invention of the streamer, the original study of angling entomology in America, introducing the dry fly to America, introducing the nymph to America, originating an American style of dry-fly tying, originating the American approach to imitation, and finally being 'the father of modern American angling traits'" (*American Fly Fishing*, 1987). While Schullery doubted that these firsts could be properly credited to Gordon, he was nonetheless appropriately respectful of Gordon's legend: "He is unrivalled in his importance as a symbol and he is rarely matched in his qualities as a writer and angling thinker. But he has unfairly been jerked from his context by our half-century binge of admiration." Knowing very little about Gordon, after reading Schullery I made three seemingly logical assumptions: first that he had enjoyed a long career, something like Skues's; second that he had traveled and fished widely, surely in England, observing other anglers while collating his experiences as his own ideas and practices began to take form; and lastly that he had written books, maybe a series like Halford's. It turns out that I was mistaken on all three counts. Gordon's story is best understood by beginning with a superficial overview of his life, looking for how and why my assumptions were so flawed.

BIOGRAPHICAL NOTES

The boyhoods of Halford and Skues were predictive of their characteristics as adults. To a large extent the same was true of Gordon. Little is known regarding his youth, but the chapter "The Quest for Theodore Gordon" in Alfred Miller's *Fishing Days, Angling Nights* (2001) is helpful. ("Sparse Grey Hackle," listed as the author, is Miller's pen name.) We know that Gordon's parents were well-to-do, socially prominent New Yorkers who moved to Pittsburgh, where Gordon was born, in 1854. After his father's death just a year later Fanny Gordon took her son back to the South, where she had family ties. The threat of Civil War led them to return to Pittsburgh in 1860 to live with her relatives, the Spencers. This family had a summer farm in Carlisle, where the teenage Theodore learned to hunt and fish with his cousins. Winters in Pittsburgh were not so pleasant. Gordon suffered from repeated upper respiratory infections and bouts of pneumonia. The consequences, both physical and emotional, were apparently lifelong. The widowed mother, naturally concerned for her only child, became overly protective, curtailing his attendance at school and even his summertime activities. According to Sparse Grey Hackle, those who knew the boy described him as sullen, frustrated, and withdrawn. Whether these health problems affected Gordon's growth or he was simply destined by nature to be a small man, he is said to have been no more than a few inches over five feet and to have weighed hardly a hundred pounds. At some unknown time he contracted tuberculosis, the disease that ultimately caused his death in 1915. In 1880 mother and son moved to Savannah. It's generally assumed that Gordon was well educated, but the particulars are unknown. Miller tells us that he worked as a bookkeeper and in some capacity in securities. Meanwhile his mother was tutoring him in the niceties of the Southern culture in which she had been raised, trying in the face of approaching poverty to maintain and augment his entrée into polite society and to introduce him to an endless succession of socially advantaged and financially secure Southern belles (*Fishless Days and Angling Nights*). Matrimony was not forthcoming.

Instead Gordon developed an intense interest in fly-fishing for both fresh and saltwater species.

By 1893 both Gordon's health and the financial well-being of his securities clients was failing, and the pair rejoined the Spencers, now living in New Jersey. Gordon worked off and on as a broker, at times receiving help from a relative of his father's, a Mr. Peck. Gordon's chronic illness, presumably tuberculosis, manifested itself through sporadic "bad spells," one of which led to his move into an outbuilding at the Pecks'. It's believed that his career as a commercial fly tyer began in this apartment and that the skill was learned from Thad Norris's book—his "book of books." He apparently achieved something of a reputation, accepting invitations to speak to club groups as early as 1897 (Austin Francis, *Catskill Rivers*, 2005). At about this time Gordon became a more and more frequent visitor to the Catskill streams until when another setback struck in 1904, he determined to undertake the open-air cure and moved permanently to the Catskills. (In its advanced stages, long before effective drugs were developed, tuberculosis was popularly treated through clean, fresh air, plenty of sunshine, as much exercise as could be tolerated and a healthy diet. The rules as set forth in Osler's *Principles and Practice of Medicine* (1923) were stringent: sleeping rooms with open windows year round and outdoor activities lasting six to eight hours, even in winter.) The ever-solicitous Fanny Gordon remained behind.

From the foregoing it's reasonable to suppose that Gordon was highly intelligent, not necessarily a man about town, a bit withdrawn, and the kind of individual who might pursue some particular interest with great intensity. What we might not have anticipated was Gordon's abilities as a writer, whether in letter form or though contributions to sporting journals. As he moved from place to place in the Catskills, a nearby post office was always a priority, and the postmaster must have needed several boxes to accommodate the replies that continually funneled in. Even during his last years a dozen letters went out or came in daily. In those days before mass media there could have been few individuals with as many contacts, whether friends or simply the readers of various journals. Exceptionally well versed in the literature, Gordon likely knew as much of English fishing as anyone could secondhand. He eagerly solicited wisdom, from Halford at first, then R. B. Marston, and later Skues. He was encouraged by Marston, editor of the *Fishing Gazette*, to submit the short pieces that became a delight to British readers. In a sense, Gordon functioned as the *Gazette's* American correspondent. Topics often featured what he believed to be comparisons between the two countries from an angler's point of view. His first contribution to the *Gazette* is dated March 1891, Savannah. Here he mentions attempts with the dry fly and suggests that our insects are likely quite different from those in English waters, richer, larger and more colourful (*The Complete Fly-Fisherman*, 1970). He wonders why fly tying hasn't caught on in America. From this period until his death Gordon was also a regular contributor to the popular American periodical *Forest and Stream*. And so, contrary to all my initial assumptions, the years of Gordon's major importance to fly fishing came near the end of his life, most of his experience was confined to Catskill Rivers, and he never wrote a book. It is these incongruities that make Gordon so interesting.

An Autobiography of Sorts

We are deeply indebted to John McDonald for editing *The Complete Fly Fisherman: The Notes and Letters of Theodore Gordon* (1970). The collection includes a preface by Arnold Gingrich followed by McDonald's introduction and prologue. McDonald correctly felt that Gordon's voluminous correspondence, coupled with his submissions to periodicals during the most critical years of his life, would allow us to understand the man. Gordon's contributions were edited only in the interest of avoiding too much repetition and "sloughing off what did not seem to me to be of general interest." Nonetheless, he left us with an opportunity to review fifty-seven pieces published in the *Gazette* between 1891 and 1915, most under the heading American Notes. Meanwhile sixty-three contributions appeared in *Forest and Stream* between 1902 and 1915. His column Little Talks About Fly Fishing was a favorite among his readers. Rather than flagging toward the end, Gordon's pen became busier as his energies failed. Between the two journals, twenty-eight of his contributions appeared in 1913 and 1914. In addition, numerous letters written to Skues and well-known Catskill anglers Roy Steenrod, Guy Jenkins, and others are reproduced. We would know even more had not the correspondence received by Gordon been lost in a fire in the little Neversink village of Liberty in 1913.

GORDON'S STYLE: A DIFFICULTY AND A DISCOVERY

On social occasions, when not on the river, Gordon enjoyed the delightful Victorian custom of afternoon tea, a pot on the stove. His writing, comfortable and conversational, has that exact feeling. You are his guest. But there's more. Gordon's manner is personal and sensitive, as you'd appreciate in a letter from a relative or close friend. I found this remarkable in that Gordon shared many of these thoughts and feelings with a general audience—strangers.

The "difficulty" is reserved for the historian attempting a biographical sketch. A project of that sort should include sorting out the most significant facets of the subject's life. For example, as a protagonist of the dry fly, Gordon's thoughts regarding the wet fly become important. However, the titles of individual pieces and letters seldom reveal much about the contents. Equally vexing, more often than not the "conversation" lacks a central theme or topic. As in talking casually with friends, subjects seem to come and go randomly. Accordingly the reader is treated to a potpourri of variably related remarks stemming from Gordon's observations and experiences, most often recent. There may be a "segue" of sorts linking one thought to the next or the transition maybe quite abrupt.

On the wet fly issue, Gordon notes: "The wet fly often kills best; in fact there are days when one may make a good a basket fishing wet, yet would have little or no success with the dry fly." That admission is repeated several times. Similarly, "We prefer the dry-fly, not because it is superior, but only because it is more interesting." So we're relieved to learn that Gordon was not a purist. But it's not until you reach the very middle of *The Complete Fly Fisherman* that the full story emerges. The following sentences come from Little Talks About Fly-Fishing, published in *Forest and Stream*, June 1907 and incidentally, the topic preceding has nothing mush to do with the wet fly. Then suddenly:

> It has been pointed out to me that wet-fly fishing, as practiced by the best American anglers, does not at all resemble "sunk fly" or "chuck and chance it" as described by many Englishmen and the school of the dry-fly generally. We fish up stream, often to rising trout, and one or more false casts are made in the air to free the fly and tackle from moisture and spread the hackle. The fly may not be dry, but it is on or very close to the surface. Rises are as distinctly seen as in dry-fly fishing and the strike follows in the instant or the trout is missed.

Norris left us with the notion that the wet fly was typically fished downstream in the traditional English manner, yet hardly a generation later Gordon calls this the "sunk fly" method and praises the wet-fly technique much as Halford did the disparaged "third method." Gordon assumes that the trout take the fly for a hatching adult or larva and these were the same patterns that others fished "sunk." Whether the "wet flies" tipped on their side rather than cocking was unimportant. Acknowledging that he was not above proselytizing, Gordon elsewhere indicated that the new wet-fly fishermen will become the very best dry-fly men should they choose to convert. Could the tiny anecdote relating the success of "whipped" flies on the Willowemoc, buried so deep in *American Angler's Book*, have such a persuasive impact? Gordon mentions the "Willowemoc story" several times, the first in a piece entitled "Dry Fly Fishing In America Before 1865" (*Fishing Gazette*, 1892). The paragraph concludes, "I know a number of Americans who fish the dry fly with perfect grace and precision, but they do not practice it exclusively or make a fetish of it." This American wet-fly technique must have stuck with Gordon, for in the very next month's *Fishing Gazette* he wrote, "Yet upstream fishing with the wet fly in low, clear water, as practiced by many of our best anglers, is not far removed from dry-fly fishing as some might imagine."

A Bit More Background

Catskill Rivers, by Austin Francis (2005) is an invaluable resource for anyone with an interest in the history of the region and not just the fly-fishing aspect. It features an array of black-and-white photos as well as maps of the major drainages. The book also provides details about Gordon that would have been hard to uncover otherwise. His home river, the Neversink, was interrupted in 1950 when a reservoir of the same name was built. The little towns of Bradley and Liberty, prominent in Gordon's life, are just to the west. The "upper" river extends from this impoundment for some nineteen miles to the very headwaters of the East and West Branches. Apparently the best sport was originally found from Claryville, a few miles upstream from the reservoir and through the area it swallowed up, some ten miles of river in all. Gordon also fished the Willowemoc, Beaverkill,

Esopus, and others, especially during his earlier years. He had visited Pennsylvania streams as well, Big Springs for one, and he understood that some resembled the English chalk streams more than did his home waters. During the summer he stayed at any of a number of Catskill boardinghouses and small hotels. Francis mentions several on the Beaverkill where Gordon and his mother would spend weeks. Gordon enjoyed much of the 1906 season on that storied river, clearly his second love. The Catskill streams were not difficult to access after the Ontario & Western railroad was built and were already crowded during the peak months. Speaking of the Beaverkill in May, Gordon wrote that there were "anglers as far as the eye could see." Since the time of Walton, great numbers of the best and wisest men have evinced a love of fly-fishing amounting to a passion, and the increase in the votaries of the sport in recent years has been in the nature of a geometrical progression. There are 300 fly fishers now where one was found fifty years ago.

BACK TO HALFORD

Halford's early books had as strong an impact upon Gordon, as Skues had felt. "The bacilli or microbe of the dry fly entered my system about the year 1889 or 1890 and the attack which followed was quite severe." When Gordon wrote to the great man in 1890, Halford's warm reply was accompanied by the gift of a set of his flies. "I can quite imagine that in some parts of your country fish could be taken with dry fly where the more usual sunk fly would be of no avail," Halford wrote (*The Complete Fly-Fisherman*, Appendix A). He suggested that Gordon send examples of prevalent American insects (no doubt mayflies) preserved in spirits to a professional tyer in Salisbury to have some imitations made. Gordon didn't take him up on the offer, but he did learn a good deal about how Halford tied his high-floating, prettily cocked creations. His admiration for Halford, Marryat, and the others knew no bounds. He called them "great past masters of the dry fly at work, men who think, talk and breathe feathers, quills, hackles and perfect dry flies, who can drive seventy-five feet of heavy line into the teeth of a gale of wind and place a tiny dun or spinner, floating cocked, a few inches above a rising trout. It must be confessed that this is the perfection of the art" (Sept. 1906).

After reading Halford, he experimented with a dry fly on the tail and a wet fly dropper, "half wet and half dry." It's hard to picture that, although Gordon caught a lot of fish on the dropper when it began to float after repeated false casts. The dry fly soon took over, fished upstream and solo in strict conformity with Halford's "rules." Gordon was far too observant and independent to adhere to the rules for long, though, noting that while limestone streams something like those in Hampshire could be found in Pennsylvania, they were a distinct minority, and concluding that if Americans waited for a rise to develop before fishing, they'd waste a lot of valuable time on the water. Gordon had developed considerable skill in stalking and spotting fish while progressing along the stream at a relaxed and contemplative pace, fishing on "spec." By 1904 a gross of patterns, mostly English, filled his jacket, and Gordon called the dry fly "the true faith."

Five years later, possibly influenced by Skues, his attitude had softened. "A Little Talk About The Dry Fly" appeared in *Forest and Stream* in 1909. He begins by admitting that he has enjoyed excellent sport with the dry fly but has "never abjured the wet fly and never expect to do so in the streams I usually fish nowadays." In 1913 Gordon reviewed Halford's *Handbook*: "Mr. Halford is very severe in his treatment of all those anglers who are not purists. He objects most decidedly to the use of wet flies and downstream fishing, upon dry-fly waters, and gives quite a long list of 'Don'ts.'" Skues's contrasting approach is then stated quite clearly and succinctly and Gordon seems ready to join Skuess' side until he says, "We quite understand Mr. Halford's position and opinions formed upon waters which command a high rental, and which are stocked, nursed into fine condition, and protected in order that they may afford the best of fishing for large trout and the very best sport with the dry fly only." This last sentence shows that Gordon was fair-minded in most matters for he personally preferred to fish public water (*Forest and Stream*, 1913). "I am sorry to say that free water is decreasing rapidly. It will soon be impossible to get a day's sport within a hundred miles of our large cities without being a member of a club or paying for the privilege in some way. I would not care to fish only in a well-stocked preserve, and as for catching trout that I had bred and raised, I fear that such sport would possess few attractions for me" (*Fishing Gazette*, 1904). "For real sport give us free water where the trout are critical, hard to please and highly valued when caught." (Gordon applauded officials in New Zealand who had mandated open space along both banks of their streams.)

Skues's Influence

Given that there's no record of further correspondence between Gordon and Halford, it's impressive that between 1905 and 1915 Gordon sent at least eighty-eight notes and letters to Skues as reproduced in *The Complete Fly Fisherman*. Skues attested that among his many correspondents none approached the American for volume. You'd think that Gordon might have developed a collateral devotion to the nymph. He didn't, despite accepting the concept. If the trout are feeding just under the surface upon immature insects, why not meet them there? He once tied some large, nymph-like flies for the high water of early season when he observed larvae rising toward the surface and even remarked to Skues that he had purchased a dozen of his nymphs intending to give them a try. But Gordon's preference was indelible. In some important ways Gordon found Skues a kindred spirit, whereas Halford was not, at least in one particular. Recalling Skues's bitter remarks concerning authorities, Gordon too had a distinct dislike for those who expressed "authoritative" opinions or certitudes based more upon self-esteem than knowledge born of actual experience. "The theorist without much experience is a dangerous person. Let us be liberal and kind to one another, trying to smother prejudice and cultivating a spirit of peace and good will among the brethren. We learn something from most anglers of real experience. It is a fascinating business." In the letter to Skues: "I wish that he could see a really fine performance with small wet flies fished upstream with the finest gut when a rise of trout is on. Mr. Halford is like many another. He has become an authority on dry-fly fishing and has been tempted in *Ethics of the Dry Fly* to speak authoritatively on a subject (wet fly fishing) of which he knows nothing" (Letter to Skues, May 1, 1913). It's my guess that Gordon had encountered certain American Halfordians. There was a class for dry fly fishers at the tournament of the exclusive Angler's Club at Harlem Mere in New York and there were a number of experienced and proficient fly-fishers in the Catskills. Some had fished in England. Schwiebert has told us that the prestigious Brooklyn Fly-Fishers left their club waters on the Broadhead in 1897 following the demise of the brook trout to relocate on the Beaverkill (*Trout*).

As a Teacher

Gordon never thought of trying to instruct nor did he necessarily consider himself particularly expert.

> The habits of trout vary so greatly in different parts of the country that one would have to be an angler of vast experience to speak with authority. We should not become hidebound or allow our minds to be filled with prejudice because we have been a fly-fisher for many years. Prejudice is the vice of little minds, and we may learn something from nearly everyone we meet. The great charm of fly-fishing is that we are always learning; no matter how long we have at it; we are constantly making some fresh discovery, picking up some new wrinkle. If we become conceited though great success, some day the trout will take us down a peg.

His were chats, not lectures.

I'd argue too that Gordon was a naturalist at heart. No doubt fishing formed the hub of his interests, but the spokes were many and varied. Articles and letters, often began with an observation pertaining to the interplay between season and weather and their effects upon wildlife, riparian and aquatic. Sitting on the bank at night, with a lantern, he enjoyed watching the wildlife. We follow the fox, hare, deer, and skunk and the Ruffled Grouse, turkey, woodcock, and songbirds. The trout, "Mr. Speckles," naturally gets a lot of attention but "Mr. Chub" is hardly ignored nor is the Black Bass. I like this:

> The silence of the snows is all over the land and the bright waters of our trout streams run almost black between icy banks. In our walks abroad we find a few living creatures—a jay perhaps, or a few, half-starved crows. Mother Nature is fast asleep, but will soon awaken and bestir herself. We do not object to the reposeful mood as we know that there is a good time coming. On a cold, bright day the wintry landscape is dazzlingly beautiful, and the air is full of ozone and tonic influence. (February 1906)

From time to time a publisher would attempt to interest Gordon in a compilation of his best pieces. In a letter to Skues he commented that he found this quite surprising: "The things are just simple little gabbles or talks. Encounters with trout, that is fish tales, serve as the vehicle for lots of angling books." The few anecdotes Gordon used were merely intended to relate an interesting event. If the reader derived something useful from the story, all the better. This one from the *Fishing Gazette*, 1913, offers several hints:

> One day last June I would have (cheerfully) paid $2 for just one more fly. I had tied only one, as I fancied it too dark for the season. And the hackles were too rare to waste. However there was a tremendous rise of these dark caddis flies. Many were hatching out, while older insects were laying eggs and doing stunts in the air over the water. All the water and air over it seemed full of excitement, and the trout were crazy. I broke my hook in extracting it from the hard roofing of a big trout's mouth, and there I was with the fish rising under my nose. I tried pattern after pattern and did kill one trout with a very dark Hare's Ear, but that was all. I put on the broken fly by way of experiment, and rose six large fish one after another.

There's a lot in this, an unanticipated hatch, the odd fly does the job with no backup, and an excellent account of selectivity.

It is not apparent that Gordon, like so many others, was a mayfly man. Caddis receive only cursory mention until here the reader is alerted to the possibility of the explosive excitement of a "caddis event." Elsewhere we learn that "gentle currents with a nice ripple" are inviting and that "lining," "drag," and a shadow-casting floating leader are best avoided. Gordon writes of shooting line, setting the hook, and playing fish off the reel, but it would take a lot of reading to put all of this into a primer.

Matching the Hatch

In his understanding of entomology, Gordon was a primitive. It had to be that way. The English had been at it for centuries. Halford and Marryat even had Martin Mosely, a professional, at their elbow. While he valued Roosevelt's work, time and again Gordon bemoaned the lack of an American reference such as Ronalds's. He certainly studied insect life and could identify the major aquatics, even sampling stomach contents and preserving naturals in alcohol. (He thought that there were two mayfly species, maybe more.) Matching the hatch, though, was mostly visual—matching the look of the natural on the water. And what's wrong with that? Despite his belief that our insects were larger and more colorful than their English cousins, Gordon recognized some resemblance and began with traditional English patterns as models, progressing to develop what became the "Catskill School" as reflected by delicate, sparsely tied mayfly imitations. Gordon was an impressionist, and in this, much like Halford, a colorist. He was convinced that "round-eyed" creatures such as trout have an acute appreciation of color combined with a defective sense of shape or form, as opposed to the almond-shaped human eye. (The contrast with Mottram is of note.) Matching the natural insect for size was also important. Perceptively, as brown trout took over the streams and fishing pressure steadily increased, Gordon recognized a need for smaller flies and better "imitations." Halls eyed hooks were a favorite. A number of anecdotes support the wisdom of approximating the naturals. One day he and a friend shared the water, each catching about thirty-five trout. His companion stuck with a Coachman, quite unlike the insects on the water, and complained that his fish were consistently small compared to Gordon's, taken on a blue dun, a much closer lookalike. "I do not advise the slavish following of the imitative theory, I only claim that on some streams (particularly where there is much still water), a copy of the natural fly upon the water will often give one a good basket of trout when all other artificial flies are nearly, if not quite, useless." Despite his preoccupation with color and size, Gordon always put presentation before pattern, giving great credit to his light Leonard rods, similar to Skues's. He recognized too that at times the trout's preference ran against the grain as when, during a heavy hatch of small duns, his imitation failed and sixty-three trout fell to a Cowdung. In general Gordon suggested that just a few, proven patterns should meet most needs, although archly remarking that a "gross or two *might* also come in handy now and then." His description and explanation of selectivity would serve perfectly today.

> Upon occasion it appears as if every fish in the river was feeding freely and is not particular as to the fly that's offered. Yet, suddenly, if there is a great hatch of one insect, the common patterns in use become of no value. One must have the colour and approximate size of the fly on the water to take trout that are rising everywhere. When there is a very heavy rise of one species of insect they often

seem to see no other, particularly if this insect continues to rise day after day. Anglers who believe that three or four patterns will do can anticipate defeat when the trout are fixated upon an insect of a particular size and shade of color.

When trout became selective, wings were found to be the least important component of a pattern, and hackle (of the correct shade) served adequately to suggest wing structure for both duns and spinners. Assuming that trout sometimes take a drowned dry fly as the adult rather than a larva, he produced the same pattern, split wings for the dry version and a single wing for the wet. However in the case of a prevalent mayfly suggested by his Quill Gordon he observed that the insect's wings erupted *below* the surface. In this case the wet fly became an emerger although the term wasn't used. Salmon patterns in smaller calibers were also a particular interest and turning that thought around, he asked, "Has anyone ever tried dry-flyfishing for salmon? Don't laugh, if you please. It is not impossible and imagine how wildly exciting it would be to have the king of fish take a floating fly!" (*Fly Fishing Gazette*, 1903). The Bumblepuppy, a red and white streamer pattern tied with long, active hackle fibers of teal flank, became a best seller, most often for lake and Black Bass fishing. The fame of this pattern grew to a point that, as Schullery notes, some called Gordon the "inventor of the streamer." John McDonald notes that famed Montana angler Dan Bailey later added a plump head of white hair to create his "Missoulian Spook" (*Quill Gordon*).

The Quill Gordon

The relative value of the products of Gordon's vise at the time is interesting. He received only eighteen dollars for three dozen Jock Scotts, a fancy salmon fly, and traded forty-five dozen of his flies for a ten-foot Leonard rod. The math works out to $270. By 1906, Mills & Sons in New York was selling lots of Gordon's flies—and plenty of Leonards to those who could afford one. (Gordon had his favorite Leonard.) I had hoped that during the process of sorting through 541 pages of articles, letters and notes, the genesis of the Quill Gordon would shine through. Instead the story is much more about a style of dry-fly construction and the creation of flies with the simple, elegant delicacy of the natural mayfly. Gordon liked neatly segmented quill bodies because they absorbed so little moisture. Sparsely applied blue dun hackle (nice and stiff for fast water) in variously adjusted colors and shades was suitable for various mayfly species. Delicate split tails (same as the hackle) fit the mayfly profile and pale wings of speckled mandarin or wood duck blended unobtrusively into the hackle. The wings were sometimes of the "rolled" type (a bunch of fine feather barbs rolled together and tied in upright as in a mayfly wing). This variation was apparently originated by Gordon and is certainly associated with the Catskill School. Gordon tied the pattern in three shades "to fit" the insect on the water. Some fifty years later Schwiebert identified one of these as *Epeorus pleuralis*, a major hatch of the early season (*Matching the Hatch*, 1955). (Angler names for the same insect included Blue Quill, Blue Dun and Gordon Quill.) Art Flick displayed photos of a Dark Quill Gordon next to an Iron frau-dator, a different species, but from the same family (*New Streamside Guide to Naturals and their Imitations*, 1965). The "Gordon," even more widely impressionistic, was tied with a gold floss, wire-ribbed body, badger hackle, and a mandarin wing. William Bayard Sturgis called it a "fancy pattern that suggests a variety of insects, sedges included, that are on the water in late May and early June. It repre-sents no individual hatch so far as is known, but is, rather, a compro-mise to be used when the prevailing 'blue tone' of the spring hatches gives way to the brown of midseason" (*Fly-Tying*, 1940). (Although Gordon depended upon his own designs, certain grand old patterns such as Wickham's Fancy remained first team players.)

THE CAHILLS

Dan Cahill was a railroad man in the Catskills dating back to 1884 and the originator of the Light and Dark Cahills, still very much in use. He and Gordon were fishing companions, and Gordon liked the impres-sionistic look and design of the patterns and tied them in the popular Catskill style, helping to bring them to the attention of the dry-fly public. As noted, the Cahills cover a multitude of hatches, Eastern and Western too. There are many variations in the styles and reci-pes for tying the Cahills, Gordon Quill, and other historic patterns—wingless, parachute, cripple, and emerger. Noting twelve variations of the Cahill, J. Edson Leonard said, "It's not the pattern, it's who made it" (*Flies*, 1950). Dan Cahill is said to have played an unusual role in bringing rainbow to the Catskills. In the 1880s a train carrying the trout wrecked. To save the fish, Cahill managed to dump the lot into nearby Calicoon Creek, a tributary of the Delaware, where they prospered (*Catskill Rivers*, 2005).

TRADING FEATHERS

Appreciative of Nature's various creatures, Gordon was also a hunter, a crack shot not above "collecting" select furs and feathers. "The most important ones [birds] are to be found at home and not in the jungles and swamps." Gaudy trout patterns were a source of irritation. "Our modern flies are mostly gay beauties, attired for conquest in the most costly stuffs, as are fashionable women at a bal masque." Skues was of a similar mind and I imagine that about half their correspondence related to tying and materials. The reading is a bit trying on two counts. We lack matching replies from Skues after the hotel fire so it's a one-sided conversation. Second, their correspondence is heavily "feathered," that is, concerned with the offerings of birds from the ever-popular starling to the Land Rail, coot, meadowlark, snipe, curlew, blackbird, and even the robin, to mention just a few. Among birds of a feather none was of more general utility than the starling, always high on Gordon's wish list when contacting Skues. Although mating pairs had been released to honeymoon in Central Park, the birds weren't yet common.

The importance attached to subtle shades of color and texture was amazing. For instance, Marston was delighted to receive the skin of a golden flicker found dead under a telephone wire just as Gordon was indebted to Skues for a gift of feathers from a field fare. Gordon worried himself almost to death (and Skues too) in finding just the right materials for the Tups Indispensible. In his last year Gordon developed a real fondness for a rooster he had acquired, both as a pet and as a source of the occasional hackle. And farsighted once again, he worked with a local farmer, attempting to breed a source of the ever-elusive perfect blue dun hackle. Many words were also allotted to Gordon's constant search for the ideal hook. He eventually had them made in England, and this brings an off-topic observation to mind: Angling writers seldom mention contemporary events, even those of considerable historical importance. As hostilities broke out between Germany and England in 1914, British ships began to fall prey to U-boats. Gordon hoped that an expected shipment of hooks would escape! In a note he sent to Roy Steenrod with a copy of *The Field*, he complained, "It is spoiled a good deal by the war news, which *properly* has no place." Skues's remarks too suggest that he found the Great War something of a nuisance. I suppose we all turn to pleasant preoccupations for relief in times of stress.

The New Century: Fisheries, Fish, and Conservation

Fishery degradation, bemoaned in Uncle Thad's time, continued as deforestation and the draining of swamp lands for development led to flooding, silting, ice jams, and scours. Smaller populations of aquatic insects, especially mayflies, and fewer and shorter hatches were the result, and caddis and small stoneflies were becoming more important to anglers. Sawmills, built as the hemlock forests were cleared, led to a secondary form of pollution. Gordon wrote of a day when the Neversink was "running more sawdust than I'd ever seen."

> How we detest a sawmill on one of our favorite streams! The sappy, heavy sawdust not only floats on the surface, but sinks to the bottom and permeates the entire river. The trout will not rise; in fact I do not believe that natural flies would be noticed, even if they would come up through the trash, and hatch out on the surface. Country sawmills usually quit at 6 PM. I wish they would quit at 6 AM and never start up again. What a lot of trees would be saved to glorify the forests. Wood has advanced so much in price that every little piece of pine or hemlock in the country is hunted out and doomed to swift destruction. They can not be replaced in a hundred years.

Gordon nonetheless commented favorably on the forthcoming damming of the lower Esopus as benefiting anglers from New York City. You wonder, though, what his reaction might have been had he foreseen his beloved Neversink under hundreds of feet of water?

The first German browns ("von Behr's trout") arrived in 1883 at Cold Spring Harbor, Long Island, with some going north to a hatchery operated by Seth Green in Rochester, New York. (Green, the father of fish culture in America, was internationally famous for his innovative techniques and is also credited with the design of the first barbless hooks.) The eggs of brown trout native to Loch Leven were received at a hatchery in Michigan two years later. The newcomers slowly began to take the place of the native brookies. Browns were stocked into the Neversink in 1886. At first Gordon bitterly resented the claim that the brookie was a char, not a true trout, and eagerly defended the natives. But over time, as the browns' dominance grew, it became apparent that they bred better, grew larger, and tolerated warmer water. An effort was made to cleanse the Willowemoc of browns and restock it with brooks. When that failed, Gordon noted that the browns were back and pronounced it "all for the good." He developed a special affection for the light-colored fish he called golden fario, which he suspected might be a species distinct from the darker German browns. More likely these attractive trout originally came from fertile lakes in Scotland such as Loch Leven where red chironimids were a major food source.

As late as the 1940s we Colorado fishermen foolishly thought we could distinguish between the lighter colored Loch Leven and darker German brown. But Behnke tells us, "The great range of variation in the appearance of brown trout resulted in about 50 species being described originally for what is now regarded, based upon genetic studies, as a single species" (*Trout and Salmon of North America*). The salmonids are among the very oldest families of fishes and their original genetic stock was scattered as continents separated and volcanic events created new mountain ranges and drainage systems. Genetic drift is a natural adaptive process in response to factors such as climate and food supply. Accordingly isolated populations became genetically dissimilar to varying degrees, differences in appearance or phenotype being one result. When genetic identities have separated beyond a certain point the members of one population can no longer interbreed with those of another and thereafter constitute separate species, e.g. browns and rainbows. (Before this process, called vicarient speciation, was discovered, it was not understood that grayling could only survive where nature had intended, much to the disappointment of Gordon and others familiar with the fishing in England.)

Stocking on the Neversink was a community effort. Beginning in 1886 farmers along the river received allotments of fry and fingerlings. Gordon's friends Bruce Leroy and Herman Christian took on the responsibility of distributing them in the stream. The stocking of rainbow was not so extensive although they provided good sport for a time on the lower Esopus. Gordon fished the Esopus and liked the rainbows' small heads, streamlined bodies, and explosive runs.

Another game fish familiar to British anglers was the chub. Gordon had mixed feelings about them. "A chub," he once said, "is better than no fish when the water is low and warm and few trout are to be enjoyed, Mr. Chub, who often pretends to be a trout, can be taken with long, accurate casts and small flies. Confidentially, I would rather catch big chub with dry fly." However he once canceled a trip because of a blossom of "chub." "They mess up the fly and waste much time," he complained. He found them good for chicken feed. Gordon also very much enjoyed fishing for Black Bass with flies, employing the same basic strategies as for trout, and worked for some time in designing a special pattern for these fish. The East Branch of the Delaware was his favorite fishery.

> During his early days Gordon records some prodigious bags. "Twenty-four large fish caught; twenty killed." His tone was more "job well done" than concern for the fishery. An opinion to the effect that large browns become a menace by cannibalizing great numbers of their smaller fellows supported the practice. He was accustomed (really almost obliged) to giving his fish away, supplying folks in and around the small hotels where he stayed. As time went on, though, Gordon's views began to take on a conservative bent: "Some say it is well to kill off big fish. I doubt this greatly. Think of the number of big, healthy ova which will hatch into large, strong fish after the mating of a single pair of four-pound fish." This sentiment notwithstanding, like the other greats Gordon constantly sought out the largest and most difficult adversaries. "I fished for one fish for more than two weeks before I got him, and had cast over his lie at least fifty times on the successful evening before he rose." Gordon was dismissive toward the smaller headwater, believing that wild streams supported so few hatches that his flies would be of no value, and had a positive dislike for showy patterns, the Royal Coachman in particular. (He saw the pattern as a poor imitation of a red ant.) He decried the sale of catches to inns and resorts as well as the practice of hiring local anglers to fill large bags for visitors to take back to the city. Some men sit in a barroom all day, after engaging a couple of local anglers to fish for them. Their ideas of what constitutes sport are peculiar, but they

usually return to the city with a large number of trout. Doubtless they enjoy a fine reputation at home. Greed and the spirit of competition should have no place on the trout stream. It is amusing however to see a number of men trying to get ahead of each other and to fish all the best water first. Take it easy, fish slowly and very probably you will have as much success as anyone.

Gordon never approved of the worm, even fished upstream, feeling that too many small fish were killed and that a few bait fishermen could spoil the fishing for many other anglers. Trout taken on bait were considered "lost." He nonetheless had friends who fished bait and when they shared the water together, Gordon was always allowed to lead the way up the stream. The minnow was regarded with the same dread. At first a champion for the common man on the stream, Gordon was a realist. Writing for the *Gazette*: "If your chalk streams were free to all, what sort of dry-fly fishing would you have? The worm, minnow and maggot would play the devil with them. If everyone is to have equal opportunities, the rank and file will have to exercise some self-restraint, which they are little inclined to do." His comments regarding Seth Green's smallest barbless hooks (made from pins) are of interest. "It is not worth while to consider the question of removing the barbs from small fly hooks. An angler who handles the small fish he catches carefully and returns them gently to the water will kill very few undersized trout."

Nevertheless, Gordon admitted to some familiar emotions regarding fishing success. Enormous bags killed with the fly were much to be admired, but Gordon expressed some degree of ambivalence when he wrote:

> I heard recently of a fine opportunity for creating a preserve—a lake and stream with much land, adjoining a State Forest reserve and another large club reserve. When our neighbors on the stream are doing nothing, our bump in self-esteem is greatly enlarged. When the other fellow turns the trick instead, it's hard to look pleasant and tender our congratulations. I rather fancy the proposition, but to make it inexpensive one would be obliged to have a fairly large membership.

(There was no mention of restrictions.) Gordon's involvement in fisheries conservation is documented by Austin Francis in *Catskill Rivers*.

In 1893, together with a half-dozen others, he helped incorporate the Fish And Game Protective Association of Sullivan County, N.Y. One goal was to recruit wardens fully deputized to arrest and prosecute violators. While there were no limits as yet, it was illegal to fish out of season, to employ prohibited methods or to kill small fish.

Reclusive by Choice or Circumstance?

Maybe it's unfair to attach this adjective to Gordon. To undertake the fresh-air cure for TB full throttle was part of Gordon's strong character. There could be no retreating to the city at the end of the season. In this March 1906 piece from the *Gazette*, he expresses hope and defiance.

> I never understood why so many people object to the country, particularly in the winter. To me it is always more cheerful than a big city. Can anything be more dismal than London or New York during the short days of winter? The light is cold and hard; even if the sun is shining one cannot see it. There is precious little real freedom in the world, and the city dweller has none at all. Perhaps I am something like an Indian in my tastes. [You can't help but recall Edward Grey's intense dislike of London summers; the two men so alike in some ways.] Strong men are sometimes alarmed by continued weakness or the loss of a little blood, and are afraid to exert themselves in any way. [Spitting up blood was common for the consumptive.] It is little use to sit about, drive a little only, and possibly eat well enough. Every organ and muscle must be strengthened, and as soon as one begins to put on some real, hard flesh, he is safe enough. It may take time, but anything is better than being a chronic invalid.

Neither did Gordon deny his loneliness and sense of desolation. Published a month later in *Forest and Stream*:

> It is a bitter cold winter's night and I am far away from the cheerful lights of town or city. The north wind is shrieking and tearing at this lonely house, like some evil demon wishful to carry it away bodily or shatter it completely. The icy breath of this demon penetrates through every chink and crevice, of which there appear to be many, and the wood-burning stove is my only companion. It is

on nights such as these, after the turn of the year, that our thoughts
stray away from the present to the time of leaf and blossom, when
birds were singing merrily and trout were rising in the pools. O
for a brother crank of the fly-fishing fraternity, one who would be
ready to listen occasionally. . . .

You can't help notice that these pieces were published consecutively,
the first, stiff upper lip for a British audience and the contrasting
second for American readers. Despite all the privation he had been
willing to endure, Gordon was an inveterate roll-your-own cigarette
smoker. It couldn't have helped. Perhaps the cigarettes became a
companion. (Incidentally, primary lung cancer was a rare disease at
the time.) Given that active tuberculosis was an indelible part of his
life, was it really necessary to hole up in the mountains, winter and
all? Roughly one in six deaths in this country in Gordon's day could
be attributed to tuberculosis. So why not turn to a sanitorium such
as the famous Saranac Lake facility with a specialized staff and programs in a communal setting? Gordon simply couldn't afford it. By
1905 there was considerable demand for his flies and contributions
to journals, an income surely, but not enough to cover recuperation
at Saranac Lake. Perhaps this makes it easier to understand why he
preferred his own place, a fine river and its environs right in his own
front yard with no rules or regulations to be followed—his own man
in his own world.

Gordon comes across as an affable enough fellow. He mentions
the pleasant occasion of an annual meeting of a sizable local anglers
club in 1904 and admits to a sense of emptiness when the guests
at local hotels return to the city at season's end. As far as sharing
the stream though, he welcomed a bare handful of friends—Herman
Christian, Steenrod, and a few others. As we noted earlier, Gordon
had developed a strong dislike for followers of Halford's purism and,
likely on this account, distanced himself from the upper-crust big-
city anglers. Bachelorhood was certainly part of the story and yet
Gordon was most respectful and appreciative of the gentler sex. "We
could never live or be happy without our greatest blessing, women."
Not known for his neatness, he did complain of their tendency to
"collect items of importance to the fisherman, box them up and
put them in obscure places." He goes on: "I am surprised that more
women do not go fishing. I find that, given the opportunity, many

women are interested in the sport, particularly fly-fishing. In fact, they are enthusiasts when they do take to it. Casting from a boat is easiest, but a woman can wade if she wants to, just as well as a man. The best chum I ever had in fishing was a girl, and she tramped just as hard and fished quite as patiently as any man I ever knew. She wore a Tam O'Shanter, sweater, short jacket and skirts, with stout shoes and leggings and waded, as I did, without waterproofs, which are only a nuisance in warm weather." Given the importance fishing held for Gordon, "best chum" says a good deal. *Catskill Rivers* displays a photo of Gordon reclining with this lady, seemingly rather uncomfortably, on a rocky bank. In another view we see her in action, rod in hand. She was not a local, and her presence was never recorded or mentioned again. For me this raised the question of whether close contact with individuals afflicted with active tuberculosis was to be avoided. Was it a stigma? It's hard to know. Osler indicates that given the frequency of the disease and the cost of institutional care, 95 percent of patients were cared for at home, practicing the "outdoor life" principles as well as they could. It did not follow that other family members would necessarily contract the disease (*The Principles and Practice of Medicine*). On the other hand, when he was tying, Gordon is said to have had the bad habit of spitting on the floor. We'd best leave it at that.

The Consumptive

Individuals with smoldering tuberculosis were called consumptives, and the term was apt in that the disease gradually consumed strength, vitality, and tissue. Bouts of secondary pneumonia often served as the final event. (It wasn't until streptomycin was developed in the 1940s, followed by other antibiotics, that tuberculosis began to be controlled.) Gordon poignantly chronicled his advancing disease, displaying courage and determination.

> If a man's lungs are in a bad state after pneumonia or any lung disease, let him hide away to mountain trout stream country where the air is dry and bracing. Let him fish and shoot just as much as his strength will permit. For a time it may be very hard work. He has no breath to go on, and puffs and blows with every little exertion, but gradually the cool, pure atmosphere will strengthen him, the

lungs will clear and heal, and in time be as good as new. One can live cheaply in the country, and can soon become accustomed to the lack of some of the comforts and luxuries of life.

A letter to a friend, Guy Jenkins, written in 1912, shows that Gordon's strength had failed badly. "The best way is to live right on the stream and it is hard to find a place where you can work, have mail facilities, and yet keep an eye on the insects and fish a little at the time of the real rise. I'm quite satisfied with two hours if it is the right time and not far to tramp." Gordon mentions a visit to a health facility in Michigan in 1913, but afterwards refused medical care. He wrote of fever, loss of weight and strength, and bouts of exhaustion. "I will be working cheerfully at something and this goneness will suddenly grab me." At the same time it's likely that Gordon's self-imposed Spartan existence prolonged his life. Despite his illness for some years he apparently felt well enough to spend long days wading the Neversink and was frequently invited to fish other Catskill streams. To have reached the age of seventy was something of a feat given that at the time, many victims of tuberculosis succumbed before middle age.

The Last Years

We are indebted to Herman Christian and Roy Steenrod, his frequent companions on the river, for most of what we know of Gordon's last years. (*Catskill Rivers* together with *Fishing Days and Angling Nights* are essential resources.) Both residents of the Neversink valley, they were rugged individualists, outdoorsmen, and colorful characters in their own right as well as expert fly fishermen and master tyers. Describing Gordon's fishing style, Steenrod admired his cautious stalking techniques and effortless casts, pinpoint accurate, and his presentation, ever delicate. A paradox: Gordon never gave up the search for still more effective flies and sought feedback from respected friends to whom he gave his experimental patterns. But at the same time he jealously guarded his tying techniques, even from Christian, who said that when a fly was in the vise when he dropped in for a visit, Gordon would quickly lay it on the table. There may have been a reason, for in Gordon's view there was no Catskill fisherman as accomplished as Christian, possibly including himself. Also a farmer, beekeeper, trapper, and hunter, Christian was a man of the soil for whom these activities came easily. He only shared his secrets with Steenrod (who later developed the famous Hendrickson pattern). Further, Steenrod was on his honor not to disclose what he had been taught. It was a curious mixture of generosity and paranoia. Later Steenrod said that while some tyers who followed may have had more skill, it was the delicacy of Gordon's flies and the exquisite attention to color detail that set him apart.

Gordon gradually became less outgoing and approachable as his disease progressed and, ironically, his fame grew. "Have you ever stayed at a place where a good many people seem interested in your sport and meet you to enquire about it? I was much annoyed by strangers, who, doing nothing themselves, came back to watch me." The loud click of his favorite reel irritated him. "I detest attracting the attention of everyone in the county when I'm fishing." Steenrod remarked: "If Gordon liked you, it was alright, but if not, you had better keep out of his way; he was kind of a cranky old cuss."

In Alfred Miller's view, Gordon's relationship with his overly protective mother cast him in a bad light. Fanny, something of a chronic invalid herself, visited for several weeks during the summers, otherwise residing with relatives in New Jersey or New York. Gordon complained: "The care of a sweet, cheerful woman in delicate health is quite an obligation." Miller observed that, "even allowing for his natural boyhood frustrations, he was an unappreciative and ungrateful son" (*Fishless Days and Angling Nights*).

Christian and Steenrod kept watch for big fish in the nearby river and would take the old master out for a shot at the prize. In his last letter, written to Skues on April 30, 1915, he discusses an insect with a "soft, buttery body" that he is attempting to imitate, indicating that he is writing from bed. Gordon died quietly the following day at the Anson Knight farm where he had boarded for the past four years. The funeral was attended by Steenrod and four of Gordon's relatives. Fanny, too ill to attend, passed away eight months later. Gordon had been assembling a series of his articles for a future book, but someone, perhaps Mrs. Christian, fearing that the pages might harbor tubercle bacilli, burned the lot immediately after his death. Skues wrote a touching good-bye, an account of a day when his only success came thanks to a pattern he had shared with Gordon: "So I owed the very last trout of a long life, with over sixty years of fly fishing, to a correspondence of some fifty years ago with a kindly American, long since dead, whom I never met" (*The Complete Fly-Fisherman*, Appendix A).

An Afterthought

I'm constantly reminded that fly fishing's greatest legends hardly penetrate the public consciousness. Out of curiosity I checked several websites devoted to the Catskills, looking for pages pertaining to the history of the region and opportunities for recreation. There's lot's to do. The visitor can hike, swim, play golf or tennis, attend shows, and gamble—entertainments also available in the city. We are treated to a long list of entertainers who performed at Grossinger's and other resorts. It's a thrill to read that Frank Sinatra sang there on several occasions and that artificial snow was first made there in 1952. But I found no mention of fishing, just as Gordon is missing from an honor roll of famous Catskillians. A website for the Neversink has a list of insect inhabitants, even a picture of a damselfly, but nothing about fishing, fly of otherwise. I realize that Gordon's Neversink lies deep beneath the waters of a reservoir and that the village of Liberty, where he sometimes stayed, became Grossinger's, but after all he is pretty universally accepted as the "father of dry fly fishing in America." I'm offended. We should all be! Fortunately, the Catskill School survived and flourished in the capable hands of Reuben Cross, the Dettes, the Darbees, and many others. ⊸►

La Branche and the Dry Fly, American Style

*B*y the early years of the new century serious American fly-fishermen had been pretty thoroughly exposed to Halfords's principles and practices. Intriguing no doubt, they described a type of fishery little known here. Thus when George M. L. La Branche's *The Dry Fly and Fast Water* came on stage in 1914, an eager audience awaited. His was a fresh approach centered upon an innovative style of casting and of dry-fly presentation ideally suited for swift mountain streams such as those in the Catskills. Due respect was paid to Halford, yet La Branche's ideas and approaches were diametrically opposed to some of the very foundations of the Halford style. La Branche's associations with other leading fly fishers, many members of the Angler's Club of New York and Edward Hewitt in particular, further assured that the book would be well received. Curiously, a fellow New Yorker, Emlyn Gill, had published *Practical Dry-Fly Fishing* two years earlier. It was thought to have adequately covered the basics, but it was mysteriously eclipsed by La Branche's book. (Copies are very rare.) According to Austin Francis the two authors knew one another and fished the same water on the Willowemoc. Gill's book even included one of La Branche's innovations (*Catskill Rivers*).

The American Dream

Born in New York City in 1863, La Branche came from a family of limited means. Joining the work force before finishing high school, he was employed in a coach factory and worked in a variety of other trades. Meanwhile La Branche maintained an improbable and precocious interest in the stock exchange, learning enough to secure a position as secretary to a prominent broker. His firm that was assigned by J. P. Morgan to market stock in the new company known as United States Steel. Successes on Wall Street culminated in the formation of La Branche & Co., a firm that was still in business at the time of La Branche's death. The frontispiece etching in La Branche's book shows him rod in hand (a 3 oz. Leonard), wearing boots but otherwise jauntily attired in a hunting coat, with a tie, winged collar, and soft-brimmed hat. Just as Halford was a key figure in the formation of the Fly Fisher's Club, so was La Branche one of the founders of The Anglers' Club of New York in 1906. His book was dedicated to the Club. He and his wife Emla were prominent figures in New York society for many years.

Halfordian for a While

A wet fly sitting the surface long enough to be taken by a rising fish was the first inspiration for many dry fly fishers, La Branche among them. The Junction Pool on the Willowemoc was home to a nice pod of surfacing trout that had persistently ignored La Branche's wet flies. Having read about trying a floating fly in this circumstance, he tied a version of a Queen-of-the-Waters with upright wings, hoping to achieve some degree of buoyancy. His first rainbow ever took the first cast. (This would have been about 1899, shortly after rainbow had been stocked experimentally.) The fly, now "bedraggled and slimy," would no longer float, but by attaching a sequence of fresh copies, he took four nice fish in short order. La Branche then, for a time, fell under Halford's spell.

> For several years after my first experience with the floating fly I used it in conjunction with the wet fly, and until I read Mr. Halford's *Dry Fly Fishing in Theory and Practice*, when recognizing his great authority and feeling that the last word had been said upon the

subject, I used the dry fly only on such water as I felt he would approve of and fished only rising fish. The dry fly fisherman has passed through all of the stages of the angler's life from the cane pole and drop line to the split bamboo and fur and feather counterfeit of the midge fly. He has experienced throes of delight each time he advanced from the lower to the higher grade of angler. Let him who doubts put aside his prejudice long enough to give the premier method fair trial, and soon he will be found applying for the highest degree of the cult—the dry-fly man.

He explains that the "approved water" suitable for the chalk-stream style was restricted to the glassy surfaces of large pools. An authority at the time even insisted that applying the dry fly to "brawling, impetuous streams of mountainous districts" amounted to an affectation. Unfortunately, La Branche's favorite streams, the Beaverkill, Neversink, Willowemoc, Esopus, and his home water, the Brodhead, fit this description. And so, fearing that he was "profaning the creed of authority and inviting the wrath of his gods upon my head," he decided to experiment regardless. The result was that after developing his fast-water technique he "abandoned the use of the wet fly for all time." La Branche was careful, though, to avoid any claim that the dry fly demonstrated superior effectiveness over the wet (fished upstream, natural drift) "except under certain conditions" and preferred it only that "greater fascination attends its use."

The Invented Hatch

Merely challenging rambunctious streams with a dry fly is not what the legend of La Branche is all about. Rather, it's the unique manner in which he attacked the "fast water" and its trout. Recall Halford's oft-repeated claim that excessive casting or "hammering" made trout "shy" and hence difficult to catch. La Branche agreed—up to a point. He put little stock in the trout's memory, but nonetheless observed that trout in heavily fished waters become highly sensitive to an angler's presence, for instance the flash of rod or line or a sloppy cast. However he further proved that, if not forewarned or actually frightened, the same fish were no less willing to come up for a properly presented fly and might even be induced to do so through multiple presentations.

I was taught to believe that if a rise was not effected on the first few casts, subsequent effort on that water was wasted—that the trout would take the fly at once or not at all. I clung to this belief for years, until one day I saw a fine fish lying in shallow water and took him after casting a dozen or more times. Since then I have taken fish after upward of fifty casts, and I rarely abandon an attempt for one I can see if I feel certain that it has not discovered me. Even when I have not actually seen a fish, but have known or believed one to be lying nearby, the practice has proven effective.

La Branche's basic instructions pertaining to a rising trout begin with an absolute and fundamental rule:

> The "swirl" or spot of a rise should *never* be cast to directly. Instead the fly should be placed a yard or two above and foot or two to the side of the current nearest the angler and allowed to drift for at least several feet below the swirl. . . . if no rise is effected he should place his fly in the same spot again and again until he has made twenty-five casts or more. It is important that each cast should be executed with the same precision and delicacy as marked the first attempt. Now here's the crux: The method is based upon the theory that a feeding trout,—or even one that is not feeding, for that matter,—may be induced to take up a position in line with the direction in which the angler's fly is traveling, under the belief that flies are coming down-stream in such quantities as to make them worth investigating. Once this position is compelled it is only a question of time and patience upon the part of the angler.

The difference between "induced" (teased) and "compelled" (commanded) is of note. La Branche acknowledged a measure of power over the trout, something we all have fervently, if futilely, desired. The entire theory of forcing the fish to rise to the fly is based upon the notion that a trout may be decoyed, provided of course, the fish is not asked to come any great distance. In one of his *Little Talks About Fly-fishing* (1907), Gordon suggests another possibility: "I know men who think the fish can be badgered into taking the fly. They go to a good pool and pound away until, as they say, a trout gets so angry that he takes the fly to get it out of the way or kill it." Do we sense some friction here? (Gordon and La Branche knew one another for some years before *The Dry Fly and Fast Water* was published.)

CHAPTER SEVENTEEN

Schwiebert jumped in too, noting that at the very beginning of a hatch, trout respond on the surface only after some time has passed, and on this basis also doubted La Branche's thesis (*Trout*). Salmon don't often respond to a hatch in the manner of trout and yet La Branche applied the "induced hatch" to salmon in just the same way. I have applied all the tricks of the trout stream to salmon fishing, with the exception of the imitation of the natural insects, and have found them useful and, in the main, successful. La Branche was convinced that salmon do not feed in fresh water and believed that instinct explained the angler's success, whether with the traditional downstream and swing-across presentation or a drifting dry fly. That instinct, he explained, was the salmon's protective urge or need to "exterminate anything that might endanger the security of the eggs to be spawned." An "induced hatch" would presumably constitute a threat of some sort (*The Salmon and the Dry Fly*). (The 1951 edition of *The Dry Fly and Fast Water* includes a second volume entitled *The Salmon and the Dry Fly*.)

The Bulger Problem

Part of a chapter is devoted to Halford's nemesis, the bulger. La Branche believed that he could distinguish between the hump or bulge made by a nymph feeder and the surface break and rings and ripples that follow a surface take. In any case, when bulgers were at play, knowing that an emergence would likely soon follow, La Branche fished "dry" to create a "hatch of frauds." He always gave the skillfully fished wet fly its due and even suggested casting a hackle fly (wet) some distance above the imagined, latest position of the bulger. When discussing glassy pools he also found the wet fly at least the equal of the dry and "more likely to be taken on the first or second attempt." But the brash American, perfectly familiar with *Minor Tactics*, said some other things about bulger fishing that might well have offended Skues. "The sunk fly method doesn't appeal to me at all. . . . in all of angling there is no greater delight than that which comes to the dry fly angler who simulates a hatch of flies and entices to the surface of the water a fish lying hidden, unseen, in the stronghold of his own selection." He referred to the induced hatch as the "highest degree of the cult." (In later years La Branche wandered from that altar to the extent of experimenting with nymphs and earned a reputation as an excellent streamer fisherman.)

The Casting Skills of a Marryat

Schwiebert called La Branche "the high priest of American presentation" in recognition of the remarkable feats he accomplished with his Leonard rods (*Trout*). (He is said to have owned thirty or more.) A cast of 117 feet set a record for the Anglers' Club of New York in 1908. Participating in a casting exhibition in England, he earned a most flattering review in the *Fishing Gazette*: "His flies go where he wishes them to go and act as he directs them when they get there." La Branche was at his best on streams where accuracy, avoidance of drag, and a friendly relationship with riparian vegetation are crucial. In tough spots with trees behind he employed the steeple cast to great effect. Otherwise he favored the horizontal cast, rod kept low, to diminish the chance of spooking the fish. When small trout were likely to be scattered during an upstream approach, La Branche positioned himself more perpendicularly, shooting sufficient slack across the current to forestall drag. Since it was his observation that mayflies most often drifted down the current, tail first, he put a nifty little downstream curve in the leader tip, placing his fly just ahead of the leader—over and over again! La Branche's signature though, "written in air," was his fluttering or bounce cast. The fly danced along the surface, alighting and lifting off in three or four sequences. Edward Hewitt described the bounce cast in *A Trout and Salmon Fisher for Seventy-five Years*, where photos show La Branche in action.

> The tackle was fairly stout, a short line with a nine foot leader and 3X tippet. Checking the rod at an angle of about forty-five degrees, the line shoots forward in the air, and as the fly nears the water, he [La Branche] raises the tip of the rod slightly and draws the fly so that it will strike the water in advance of the leader; in fact he often keeps the leader out of the water altogether. By this means the fly often makes several skips on the water before it comes to rest.

It was a difficult trick, best practiced across stream and sometimes greatly aided by a strong upstream wind. As we shall see, La Branche's idea of activating an artificial, giving it movement as an inducement to reluctant trout, later influenced his friend Edward Hewitt's strategies. La Branche was hardly the first to show that introducing a spark of animation in an otherwise sedentary imitation might arouse the predatory instincts of nearby trout. Dappers and dibblers had always known that, but they were not actually casting the fly.

Caddis Capers

The bounce cast, although not La Branche's standard, would seem a perfect imitation when caddis were skating or fluttering on the surface. If so La Branche made no mention. Like his predecessors, he was a mayfly man. It occurred to me that once again the role of the caddis remained somewhat ambiguous. Admittedly the Grannom hatch had been anticipated since Dame Juliana's time. Pritt (and perhaps Stewart) documented the importance of caddis in North Country fishing, and Caddis patterns such as the Cinnamon Sedge and Silverhorns go back to Ronalds and made their way across the ocean in recognition of similar species here. But when Halford and Gordon recount their experiences, intending to imitate the mayfly, did the trout always take them as such when caddis were about?

This issue reminded me of a favorite book. Leonard Wright's *Fishing the Dry Fly as a Living Insect* was first published in 1972 and reprinted under the title *Flutter, Skitter and Skim* in 2001 after selling some 25,000 copies. Wright's book was one of the first to mention caddis imitation. He correctly remarked that early writers "seldom mentioned whether the fly they were describing was a caddis, mayfly or stonefly" and that, in a sense "they all forgot the caddis." Despite heavy caddis emergences on the Catskill streams, Americans were slow to catch on. Wright nonetheless believed that caddis played a significant role in early American fly fishing in that many standard wet-fly patterns made good imitations and given that droppers were often bounced along the surface to good effect. Wright brought up interesting data based upon studies of stomach contents suggesting that, over time, caddis have become more important to the flyfisher, and mayflies less so. (The mayfly is known to be environmentally fragile and is used to document the effects of pollutants such as heavy metals.) His book taught us how to tie more effective and realistic imitations that could be activated on the surface, caddis-style, without drowning. And we learned to give the fly life, presented downstream. The "sudden inch" comes off a downstream mend followed by a twitch that nudges the fly upstream, just like certain caddis in the process of shedding the shuck. The fluttering dry fly was another technique.

The Pink Lady and Un-matching the Hatch

La Branche had no pretense to entomological expertise and was doubtful regarding efforts at close imitation. Like others before him he was a firm advocate of "presentation over pattern," recounting instances when he was able to encourage a feeding fish to swallow naturally drifting pine needles or twigs he'd dropped upstream into the appropriate current seam. In discussing the attributes of a successful fly, he began with its position and action on the water, (accurate placement without drag) followed by size, form, and lastly, color. There was nonetheless some uncertainty over the issue of color. "It seems to me that the colourist, as a rule, is much too certain that his flies appear to the trout as they do to his own sense of sight." But then he suggested that an imitation of the hue of the natural fly is desirable.

The Pink Lady is a case in point. He had purchased some King-of-the-Waters tied with a bright red silk body. In the water the body became a dingy brown, and the flies were refused. But when the dye leached out overnight, a bright pink body emerged that became quite bright when wet. The contrasting ginger and light blue dun hackle and tails meanwhile suggested the hues of a number of mayflies, and he enjoyed great success. La Branche nonetheless attributed the pattern's success to its form, not the blushing body. (The Pink Lady continued to be a standard pattern well into the 1940s.)

La Branche disparaged gaudy patterns that "only intrigue foolish fish." "There is no place for the Royal Coachman in my book," he declared, and he felt the same about the popular Parmacheene Belle and infamous Alexandra, a pattern Halford called a "scourge" that should be outlawed. Both argued that the best sportsmen and best trout should shun fancy flies; "the angler who captures a 'fool fish' attains no honor." There was a suggestion, too, that these undesirables were mostly effective on wilderness streams in accordance with Stewart's view. Six patterns covered most of La Branche's needs. The Whirling Dun, Pale Evening Dun, Flight's Fancy, and Willow Fly (or versions) went back to Ronalds and were of course originally fished wet. La Branche tied his to float "in accordance with my practice," meaning upright, starling wings (even cocked a bit forward), good stiff "legs" (hackle) and tails and no doubt a lighter, eyed hook. While he admitted that hackle patterns performed as well as the winged, he liked the look of the prominent wings on the water. They were fragile, but he was willing to change to a fresh fly after each fish. La Branche liked both dubbed and floss bodies and palmered the hackle of his flies to the bend of the hook, then clipped off the body fibers nearly flush, leaving a "fuzz." It was an innovative touch that created a hint of translucency and improved buoyancy. A bit sarcastically, he commented that the only virtue of quill to suggest body segmentation preferred by Gordon was the "pleasing effect upon the artistic eye." In spite of this remark, he is known to have purchased flies from Gordon.

The smutting rise was as problematical as on English streams and greatly frustrating. Stomachs contained balls of the tiny insects and yet they flew in swarms just above the surface. La Branche wondered if the splash of a trout's tail served to drown a cluster of midges, conveniently collecting them in the film. Given that hooks small enough for imitating the "curse" were not available, he gave up, recommending that any dull-colored fly, size 12 or 14, was a likely as anything tied smaller. (At the time, #16 hooks were the smallest he used.)

STREAM CRAFT

In many ways *The Dry Fly and Fast Water* served as a primer for the fisherman working freestone streams. La Branche loved the pocket water best. Perhaps no water on our American streams appeals more to the average angler than a beautiful pool, and yet rarely does this water fulfill the promise it seems to hold out. A pocket might be found on the upstream face of a boulder, beside or below it; the same for a snag. The base of a waterfall or current strip beneath a bank would also qualify. La Branche relished small volume, high probability, current-defined targets that he could fish (and re-fish) with ultimate precision. He repeatedly used *pocket* in his descriptions, and it's very possible that it was he who introduced the popular descriptive into our vocabulary.

A chapter is devoted to the importance of compulsive observation, no feature or event too small. Look with eyes that see. Accepting the dry fly as both sport and art, La Branche criticized the "casual advocate who fails to assess and improve his abilities, among them, observation." "The careless angler frequently overlooks incidents, or looks upon them as merely trivial, from which he might learn much if he would but realize their meaning at the time." He offered an interesting observation that in the process of watching a floating fly's progress along its drift, an angler should at the same time see through the surface. "In this the mid-water and streambed are disclosed where fish might be located, lying in position, cruising or merely 'at leisure.'" His implication was that wet fly fishermen, less attuned, read and cover the water in a more aggressive and mechanical manner, using "as much stealth as someone escaping from a burning building," and is so doing panic more fish. The reader is advised, paraphrasing Walton, to "study to be quiet." All motions should be slow. The cast, often no longer than thirty-five to forty feet, should be "fished out," that is allowed to drift well downstream before being picked off the water since fish sometimes back downstream under the fly before taking. Fresh water should be progressively sampled as one moves upstream, all the while aware of small fish that may be frightened, fleeing on ahead. La Branche was tough and truculent too. Heavy wind was not considered a problem. "If you can stand the cold of a northern, watch for terrestrials that get blown into the water." Thunderstorms? The fish don't necessarily quit so why should you?

CONSERVATION

La Branche agreed with Gordon's assessment of the damage resulting from denuding the forests and also that the native brook trout must yield first place to its European contemporary, "he having been endowed by nature with a constitution fitted to contend against existing conditions and survive." Browns were also found to be the better fast-water fish. Although it seemed unlikely that either kept records as compulsively as their English contemporaries, both despised the pot hunter and the practice of turning large bags into bragging rights: "men calling themselves anglers—save the mark—who limit the number of fish to the capacity of creel and pockets." La Branche also believed the law defining legal size to be "wicked," poking fun at the angler who captures a small trout and after attempts to stretch it out to six inches fail, throws it on the bank.

CHAPTER SEVENTEEN

La Branche and Gordon: A Strained Relationship

Gordon attended a luncheon in the city with Hewitt and La Branche only a year before his death. How Gordon was convinced to attend isn't clear. He and La Branche had corresponded and even met three times, twice to fish, prior to 1914. I was puzzled to find no mention of Gordon in *The Dry Fly and Fast Water*. It turns out that Gordon wrote a mixed review of the book for the *Fishing Gazette* (June 1914) wherein the major criticism was La Branche's lack of attention to entomology or attempts at imitation and his dependence upon so few patterns. He nonetheless concluded, "I do not think that I have done the author justice." On an earlier occasion Gordon told La Branche face to face that the dry fly should be reserved for calm waters and rising fish, and fishing it otherwise would belittle the practice. He further agreed that it was an affectation to fish the dry fly on fast water. La Branche wrote, "I was upset more than a little, but persevered with my idea" (Schullery, *American Fly Fishing*). The notion that Gordon had become something of a curmudgeon is suggested by a note to Skues regarding La Branche: "Manual dexterity is his chief pleasure." (I assume he referred to the bounce cast.) Gordon bristled a bit in a letter to Steenrod: "La Branche and Hewitt declare that they would be content to go through a season with *one* fly. And that is allright, probably, *where they fish*." La Branche was forgiving of Gordon in this eulogy letter quoted by Francis:

> The greatest student of fly-fishing in this country, and without exception the best flytier I have ever known. In the great beyond there must be some little corner where anglers are tucked away, where one may hope to meet his streamside associates, and discuss again the old, old theories and experiences—at any rate, I believe there is such a place, that my friend Gordon is there now, and is rigidly reserving a seat next to his for me. (*Catskill Rivers*)

Comparing Gordon with La Branche, Francis makes the interesting comment that Gordon saw fly-fishing as a recreation and cared little for fame, even resenting the attention he received. La Branche had a different philosophy. He sought fame, and he won it for his abilities as a dry-fly angler.

In addition to diverse salmon rivers, the robust La Branche later fished the Test and River Dee in Wales and, more remarkably, now very much a senior citizen, became a pioneer in saltwater-flats fishing where he and the baseball icon Ted Williams became friends. Some credit La Branche with tying the first bonefish fly. He passed away at eighty-six after a long and multifaceted career, survived by his wife and five children.

LOOKING BACK

Gordon, perhaps unwillingly, handed the dry-fly baton to La Branche, who, with Hewitt and others, went on to expand its role. La Branche hoodwinked, finagled, or otherwise seduced many a fine trout, no longer able to resist his repetitious, perfect presentations. Edward Hewitt further attested to the magic of his method. But given the dedication, concentration, and patience the technique demands, not to mention exquisite casting expertise, I wonder how many others acquired the skill at a level equaling La Branche's, at the time or since? For a lot of us, I suspect, after a half-dozen unrequited casts, there's a strong urge to change the fly or move on to greener pastures. But there are all sorts of skillful pocket-water fishermen today.

Not to be overlooked is La Branche's wonderful capacity for packing a variety of practical observations and suggestions into a short paragraph written in a clear and pleasant manner. The text is so rich in these gems that some are liable to slip by unless you pay careful attention. I first read *The Dry Fly and Fast Water* in 1951, the fourth printing, making notes for future reference. When I go back to it I find it full of essential concepts. Maybe they're old hat now, but if so why was the book deemed worthy of a fresh printing in 2002?

Edward Ringwood Hewitt, Fly Fisher Extraordinaire

*E*rnest Schwiebert referred to Gordon, La Branche, and Hewitt, as The Trinity. Whereas Gordon is credited with the evolution of the Catskill dry fly and La Branche with a technique of presentation suited for our swift streams, Hewitt's accomplishments were too varied to label. He was interested in, held opinions about, and initiated experiments or innovations regarding practically every nook and cranny of the sport. Additionally, he outlived most of his generation, still spry and keen of mind. He was eighty-one when *A Trout and Salmon Fisherman for Seventy-Five Years* was published in 1948, ten years before his death! Dip into Hewitt's family history and you'll soon be tangled in a web of Americana beginning in postcolonial days and continuing to the present. This background has relatively little to do with fishing, yet it's crucial to an understanding of this remarkable angler.

Ringwood

The story began when rich and extensive iron deposits were discovered in northern New Jersey. Smelters developed in the 1770s were long associated with the name Ringwood, a local river important to the industry. Edward's grandfather, Peter Cooper, the son of Dutch immigrants, parlayed successful business ventures in New York City into a major iron manufacturing company in the area in 1854. His partner, Abram Hewitt, married Sarah, one of Cooper's daughters.

Cooper and Hewitt later built Ringwood, the family mansion and estate where Edward was born in 1866. A friend and associate of Andrew Carnegie, Cooper was also a powerful industrialist and philanthropist. The Cooper Union for the Advancement of Sciences continues as an important institution in lower Manhattan, and the Cooper-Hewitt Museum is a branch of the Smithsonian. Grandfather Cooper was also an inventor. His projects ranged from the practical—a motorized musical cradle (patented)—to a grand scheme for obtaining energy from ocean waves. He had a glue business, was president of the company that laid the Atlantic Cable, invented Jello, and ran for president of the United States. The Hewitts were an impressive lot, too, and one of the wealthiest and most influential families of the time. Indeed Abram, as well as his uncle, served as mayor of New York City. In short, both the Coopers and Hewitts were talented, productive, and famous. As we'll see, while Edward was never a politician, he inherited many of his grandfather's other interests and capabilities.

Edward was "to the manor born" and in a sense, also to sport fishing. The Ringwood River ran through the estate replete with a hatchery, which he and his brother, Cooper, operated as teenagers. (Ringwood Manor, all fifty-one rooms and twenty-four fireplaces, was given to the State of New Jersey in 1936 as a National Historic Monument.) Hewitt enjoyed (and utilized) an extensive and eclectic education obtained at a number of institutions, Princeton and Stanford among them, where he studied engineering as well as the biological sciences. He knew the New England streams well, often fished in Canada, and even explored rivers in the Rockies. Having ample funds, Hewitt crossed the Atlantic regularly, pursuing trout and salmon in England, Germany, France, and Scotland. When easy access by the automobile threatened to crowd the Catskill fisheries, he noted that the Neversink lacked a good road along its course and in 1918 purchased 2,700 acres including the Big Bend segment, some five miles of river. It was not far from Gordon's home water. By 1922 when his first book, *Secrets of the Salmon*, appeared, Hewitt had accumulated several lifetimes of angling experiences.

IN PERSON

Cultured, likely pampered, and intellectually gifted, what was Hewitt like in person? Schullery remarked upon his "opinionated crankiness" and Schwiebert mentions his "posture of complete expertise in all facets of the subject" (*Trout*, 1978). You might have anticipated comments such as these. Gingrich was a member of the Big Bend Club that leased Hewitt's water and was privileged to know him during his last years. He too remarked upon Hewitt's annoying degree of self-assurance and pertinacity. Whereas Marryat could discourse engagingly on matters cosmopolitan, conversations with Hewitt tended to be more one-sided and sometimes boring. Gingrich and a friend were once treated to a lengthy discussion of lecithin, an important chemical substance with many biological functions. Anyone with an interest in the health and well being of cell membranes or neurotransmission or even the formation of the mundane gallstone, would know a lot about lecithin. But Gingrich became pretty fidgety. Unable to break away: "You had to admonish yourself, as you would a restless child, to shut up and listen, reminding yourself of what a privilege it was to be talking to one of the angling great" (*The Fishing in Print*). Given this introduction I rather expected a dry and patronizing, if not imperious style of writing. It wasn't so.

The first chapter of *Telling on the Trout* (1926), begins: "My conception of a good fisherman is one that can adapt himself to any conditions and catch trout no matter how hard they may be to take. I am not one of those who hold up their hands in holy terror at bait fishing when it is necessary to get fish by this method. I do it occasionally, and try to do it as skillfully as possible." Three concise declarative sentences devoid of long words, flowery syntax, or ambiguities tell the reader that Hewitt is a pragmatist and the antithesis of a Halford. (Hewitt had read his books and occasionally refers to him as Hallford.) As a youngster he fished a brook trout's colorful fin on a long line downstream. He claimed that when the bait was "moved" in the correct manner, he might have caught all the fish in the brook. Here, early on, Hewitt developed one of his cardinal principles: In attracting fish movement is more important than any other feature, whether with fly or bait. When the ever-increasing brown trout proved difficult to catch "downstream" he wrote: "Only a few fishermen fish wet fly upstream. Those who have become skillful at it can catch all the trout they want, but they are few indeed. This is the acme of the wet fly art." When a fly needed to be fished deep, he'd wrap on a little fuse wire. Unsporting? Not in Hewitt's view—do what's necessary! His first attempts with floating flies must have been in the 1890s, and soon after he became a dry-fly man by preference. He reiterates, "Let any dry fly fisherman who has learned the easiest method at trout fishing, fish alongside a skilled wet fly man for a day, and he will find he knows little about the real art of catching trout" (*Telling on the Trout*).

FRIENDS AND ASSOCIATES

Some of Hewitt's friends were experts such as Herman Christian and Jim Payne, famed rod builder. Many were "blue bloods," George La Branche, John Alden Knight (of Solunar Tables fame), the well-known painter John Atherton, and captains of industry including Andrew Carnegie and Ambrose Monell. *Hewitt's Handbook of Fly Fishing* (1933) is dedicated "To the Angler's Club Lunch Table where all important trout questions are daily settled and unsettled." Hewitt sat at the head of that prestigious table for many years. But although most of his intimates came from angling's high society, Hewitt clearly felt a kinship to fellow anglers that precluded snobbery. It was skill that he admired.

I was surprised to learn that Hewitt delighted in teaching ladies the art. His daughter Candace became quite expert. Grace Vanderbilt was a student, as was Atherton's wife, "Max." The Athertons and Hewitt often fished together (*Catskill Rivers*). His *Handbook* was certainly designed for a general readership and covered everything from tackle to technique, concluding with a chapter on cooking trout, all in just over 100 pocket-size pages. (The subtitle, *Wade Mecum*, puns on Pulman's *Vade Mecum*.) An argument is made for paying no more than necessary to obtain dependable tackle, and there are other homely pieces of advice. "I have often forgotten money when on a trip, and if I have some in my leader book I can always get crackers and cheese at a country store or a ride home." Imagine a hitch-hiking multimillionaire with wet boots packed in his fish bag! A more somber warning to the effect that kidney disease is caused by fishing in cold water with wet feet reminded me of my grandfather's similar concerns.

PERFECTIONIST AND PROMOTER

Hewitt recognized three stages in a fly-fisher's maturation: the desire for the most fish, then the largest, and finally "when he studies to catch the most difficult fish he can find, requiring the greatest skill and most refined tackle, caring more for the sport than the fish."

Hewitt graded his own prowess by playing a "game of solitaire." "My temper ebbs and flows with the results. When the great day comes . when 90% to 95% of the fish I see rising are raised and hooked, I feel like the golfer who has beaten the best previous score."

Hewitt was anxious to share whatever he might have learned, but he could be impatient and brusque when tutoring others on the stream.

> A friend and experienced dry fly fisherman was having little success in fishing the difficult Neversink pools. I told him the pool had not been fished in the best way, and he should have taken three or four fish. At this he seemed greatly surprised and rather hurt. No doubt I was quite rude, as I often am, but this was the only way to get him to see that there was more to this kind of fishing than merely being able to cast well in likely places. (*Telling on the Trout*)

Although a perfectionist, planning and executing his attack with the greatest precision, Hewitt also had a refreshing ability to tell stories about or on himself. Some of my favorites come from *A Trout and Salmon Fisherman for Seventy-five Years*. There was something of the show-off in Hewitt. While still a youngster he gained notoriety by taking the only salmon of the season from a pool under a famous waterfall. Futilely attempting to leap the falls, the fish often fell back onto a flat rock at the base where they lay temporarily stunned. Our hero waited until the tourists left, then cautiously worked his way within reach of the rock and impishly hooked his fly into the jaw of an incapacitated salmon. When an audience again collected, he battled the rested fish amid great applause.

During a visit to Andrew Carnegie's Skibo Castle in Scotland, Hewitt and his father were invited to a picnic on the moors. The host had hoped to provide fresh trout from the small streams, but found the fish "dour." After asking permission, Hewitt "guddled" (tickled) as many trout from beneath the banks as were needed. Whatever it takes!

While recuperating from a throat infection at a German spa, he managed to catch a record barbel (a bottom-feeding fish) on a clump of worms he'd dug from the garden. Military officers arrived promptly, ordering Hewitt to put the barbell, still alive, into a fountain. The next morning the same officers came to take him away, he imagined to prison. Instead he was introduced to Kaiser William I who had enjoyed the barbel so much for his breakfast that he commissioned

Hewitt to catch more. They became friends, walking together, and sometimes with Bismarck, "the most forceful and sinister man I ever met." The precocious Hewitt was only seventeen!

Some stories had nothing to do with fishing. In 1891, when he was a student at the University of Berlin, Hewitt went to visit his sister in the south of France. While taking a scenic photo of a church on the coast, he was arrested and accused of being a German spy planning a location for a gun emplacement. (Feelings against the Germans still ran high after the stinging outcome of the Franco-Prussian war in 1870.) When local officials prevented Hewitt from contacting American authorities, he was transferred from police station to prison to a solitary dungeon reminiscent of *The Count of Monte Cristo*. Along the way he met some amusing cellmates, murderers and the like, before finally convincing higher ups that he was an "American prince"! (Not far from the truth.) Eventually the American Minister at Paris (who knew the family well) arranged for his freedom.

THE NEVERSINK

Hewitt's Big Bend water, elevation 1,400 feet, stayed cool during the hottest summer days. The Neversink's cold, clear water was not entirely a blessing, as Austin Francis commented, noting that it was "too cold and too pure for an ideal trout stream" and remarking as well upon the relative acidity of the river, in contrast to a lush chalk stream. On the other hand, the river was free of chub (*Catskill Rivers*). (Hewitt also owned considerable property on the Willowemoc, which he eventually sold, although reserving full fishing rights.) The boulder-strewn Neversink offered a series of large pools separated by stretches of pockets and rapids. There were shoal flats too. Hewitt noted, "The change in the style of fishing is a great rest and adds much to the variety of the day's sport." He and friends traversed his water over a rough track at frightening speeds in an ancient Buick known as the Mechanical Goat. Holes had been cut in the canvas roof to accommodate rods. A camp was constructed hard by the water, together with a sizable hatchery, and high on the ridge above he built a home for family and visitors. While Halford and Skues had small areas set aside for insect study, Hewitt's camp housed a much larger and more sophisticated laboratory supported by a fully equipped machine shop. This is where he attacked a wide variety of "problems," for beyond all else, Hewitt was a problem solver. His solutions to catching the difficult trout in his pools and flats largely came from this laboratory. But before getting into those specifics, we should understand how much science Hewitt really knew and how he applied his knowledge in practical ways.

THE HATCHERY

Well before his book *Better Trout Streams* was published in 1931, Hewitt was widely accepted as a professional fisheries biologist by his many clients. "I feel that in focusing public attention on the environment of trout in streams, I performed a real service to my fellow fishermen. I expect I know more about trout than anyone else in the county," he wrote modestly. Consulting fees of $100 per day were collected for his services from private water owners and other hatcheries. (I had assumed that Hewitt's pockets were bottomless, but Francis tells us that rather than "pursuing a family tradition of investments and the corporate directorships, Ed Hewitt dedicated his inheritance and considerable talent to a lifetime of making better trout fishing," *Catskill Rivers*.) Among his innovative ideas was the practice of stringing electric lights just above the water to attract insects at night for the "slashing fish." Predators such as kingfishers, herons, and cranes were caught in traps suspended by wires strung across the water. More scientifically, there were instruments for measuring current velocity and oxygen content as influenced by temperature, turbulence, and other factors. The Neversink flats held few fish until Hewitt built low dams made of boards to create deeper, cooler, well-oxygenated water on the downstream side where large trout accumulated.

He could also determine relative acidity v. alkalinity and water "ionization." Observing that trout feed actively when rain enters the stream and noting that rainwater carries a positive charge, he obtained a small quantity of radium for one of his hatchery tanks. The fish died. The experiment sounded ridiculous until I remembered several "radium hot springs" in the Colorado mountains recommended for sufferers from rheumatism and other chronic miseries.

Liver had been the food of choice for hatcheries until it became popular as an "anti-anemic" and hence too costly. Other ground organ meats were substituted, but Hewitt was determined to develop a dry food. Early experiments failed when the trout sickened and failed to thrive. Hewitt's microscope revealed that fat had accumulated in the trout's liver cells, and in the biochemistry section of his lab, abnormal ratios of fatty acids, cholesterol and, sure enough, lecithin, were found. The formula was improved. Anticipating Linus Pauling, Hewitt believed Vitamin C to be essential together with a certain amount of fat. He postulated that the trout's fondness for mayflies should have a dietary basis and when fat content was measured, they turned out to be quite buttery. Hewitt also suggested that formic acid is the spice that attracts trout to seek out ants, this after some flew into his mouth during a hatch. The rings formed in scales of trout and salmon proved to be a means of determining their age and well-being. A distinctive ring marked each spawning period, and the spacing reflected the relative abundance of food during a season. Hewitt always believed that stream-bred trout were superior, but when stocking became necessary he favored fry over fingerlings, admitting that predation left only a small proportion to mature, and he strongly discouraged dumping pan-size fish for small boys to "catch out within a few days by whatever means."

Hewitt on the Trout's Senses

His opinions were strongly held if sometimes unconventional. Refuting Ronalds's experiment, he reported that rising fish could be put down by loud conversation. Indicative of a well-developed sense of smell, his hatchery fish would come from some distance to a diluted feeding of lights. (Some New Zealanders believe that trout are offended by dissolved human scents drifting downstream from quite some distance.) Some of Hewitt's other ideas were based upon mere supposition. Why do trout get out of sorts when the barometer falls? Answer: Their swim bladders expand, pushing upon various organs and making them uncomfortable! Hewitt's ideas about the trout's sharp appreciation of color came from studies by German biologists. The same studies convinced him that trout can distinguish between sizes and shapes but are able to react to only one of the three qualities at a time. On this account, when trout turned away from a particular pattern, he first changed to a smaller hook caliber, a common practice still. Despite Ronalds's and Mottram's earlier studies of the trout's "window" he said, "I never like to accept a scientific fact as really so if I can possibly repeat the experiment myself. In scientific work I have often been misled." And so he invented a water tank with a tilted window through which the meniscus could be photographed with a movie camera. Another device measured the degree of light penetration through the water based upon the angle of incidence. (Working through Hewitt's explanations, diagrams, and photos becomes a bit confusing, and referring back to Mottram, I didn't find a lot that was new.) He did conclude that a trout's ability to see in low light was far superior to ours, consistent with active feeding continuing late into the gathering dusk. On the other hand, he concluded that the round-eyed trout, lacking eyebrows and lashes and with pupils less capable of narrowing in bright light, are more easily blinded by the sun. The corollary: when possible it's best to have the sun at your back.

Hewitt's most important observation became key to his most famous patterns: "Flies just resting on the surface of the water on their hackle tips depress the film by their weight where the hackles touch. These depressions act as lenses or light condensers, forming brilliant light spots, as seen from below. It is this light flash which gives notice to the trout of the arrival of an insect on the surface." In a series of black-and-white plates in the book, a grasshopper is shown, first resting quietly in the film as a dark silhouette. The next frames show dramatic light patterns when the insect begins to kick. Hewitt concludes, prophetically as we shall see, "When a natural insect alights on the water it makes light flashes exactly similar to those made by an artificial fly when the fly is properly handled." The downside was noted, too, in that a fly striking the water forcefully would create a frightening explosion of light. He found that a floating leader also acted as light condenser, creating an unwelcome, bright streak (A *Trout and Salmon Fisherman for Seventy-five Years*).

The Problem of the Pools and Flats

In common with the stretches of flat water with their low, log dams, the pools held sizable trout, secure, self-sufficient, and wary. They *needed* catching. Two impediments stood in the way. First, the fish were easily alerted to an angler's presence. Hewitt began his campaign by honoring Cotton's "fine and far off" edict. He once won a casting competition in England and as a veteran salmon fisherman, he did not consider distance a problem. Fifty-foot casts were about the norm. But even when leader and fly settled safely onto the surface, there was no guarantee of a response, even if a rise was on.

Always fastidious when matching the elements of his tackle train to best fit the situation at hand, Hewitt rigged up a twenty-to-twenty-five-foot leader with six to eight feet of 6X tippet. (I was surprised to learn that fresh, top quality 6X gut tested out at as much as one and a half pounds.) But what about the flashy leader? Back in the lab Hewitt soaked the gut in silver nitrate, exposed it to the sun, and then "developed" the leader with the same chemicals used for film or paper. He believed that tiny adherent particles of metal absorbed shine and made the material stronger at the same time. He claimed that an angler fishing his silvered leaders could expect a 30 percent larger catch. His patented leaders were sold by William Mills in New York and C. Farlow & Co. in London for many years.

Birth of the "Skater"

The trout still sometimes bolted even with a long, delicate cast and invisible leader, as if the fly itself were the problem. A smaller caliber hook would help, and so would reducing the fly's bulk and weight. Like La Branche, Hewitt doubted that trout were much influenced by a fly's wings and that hackle fibers "passed" as wings. Further, thin tippets tended to twist about the wings. Hewitt's thinking may have proceeded something like this: A wingless fly doesn't require a tail for balance, and if the hackles are good and stiff, holding the hook's shank above the surface, does a body really matter? Besides, leaving the shank bare would reduce a water-absorbing component. "I went back to my camp and tied several flies, finally making what is known now as the Neversink Skater. It was tied sparsely because I wanted it to cast easily and not have too much air resistance. It was made as large in diameter as possible, with the longest, stiffest hackles I had." (The hook was a #16, while the outside diameter of the hackles reached two inches.) Casting forty feet of line, his skater "could be made to alight on the water like a feather."

The problem of presentation was now in hand. But how to attract a response? He recognized that the realistic sparks of light in the surface film created by those long, sparse hackle points was a logical start. Two other observations contributed to the Skater's success. Earlier, while working over one of his pools, he observed, "When the trout did not rise at once all that was necessary was to let the fly rest a moment and then pull it slightly." This almost always attracted a solid rise. The next day he visited another pool where he had never taken many fish. The new twitch technique led to the extraction of fifty-two trout measuring up to eighteen inches—likely most of the population. (One was saved for lunch.) The same ploy is mentioned time and again—rest, then just a little pull.

Hewitt had also seen large trout, indifferent to every pattern, leap excitedly for butterflies fluttering above the water and fleetingly touching the surface. Since the trout were rarely if ever successful in catching the insects, it must be the movement that attracted the fish. If a mere pull or twitch could trigger a take, the long-legged thistle-down Skater could be induced to perform all sorts of acrobatics. He tested his prototype under the most difficult conditions, mid-summer and mid-day under a bright sun. When the fly had made only two jumps over the water, a large trout of about four pounds leaped out of the water right over the fly like a porpoise and missed it entirely. When he subdued the fly's antics somewhat the new "Skater" began to hook fish. He was joined by his son Abram, and the two were able to catch "six large trout right in the middle of the day, and hooked one more."

The Skater was later proven equally effective in attracting big fish from beneath dams in the flats during the evening hatch and rise. When driven beneath overhanging vegetation the fly came down gently and could be jumped, bounced and dragged, eventually driving the big browns wild. The induced pique didn't always result in a hookup, and yet as long as the fish wasn't pricked, it would continue to rise. Aroused, it might then succumb to a standard pattern.

Hewitt summarizes Skater fishing: "The action of the fly alighting on the water and while it remains floating is far more critical than anything else." How much of this insight came from La Branche's technique we'll never know, but Hewitt wrote, "This explains the great effectiveness of Mr. La Branche's 'bump cast' by which he makes the fly strike the water and make several little jumps before it comes to rest and floats into the fish's window." Referring to surface film light patterns: "It is the skill with which the fisherman imitates these light flashes of natural insects which determines the number of trout he will catch. I therefore regard the manipulation of the fly on the surface as more important for successful trout fishing than the exact imitation of the insect." Where the fly alights with reference to a diffident fish—to one side, directly ahead or even behind—could also be critical. He often dropped his fly slightly behind a fish, hoping it would turn toward the source of the flash reflected from the bed.

John Alden Knight was another Big Bend regular who developed something of a pre-occupation with Skaters. Despite the excitement these flies caused, hookups weren't so common. Knight nonetheless used this feature as a means of locating the residence of really big fish that might later be taken with more conventional patterns. (Incidentally, Hewitt thought that his friend's *Solunar Tables* were so much bunk—and likely told him so.) Hewitt concluded, "I had developed a way of dry fly fishing which would raise and hook large trout in all kinds of water at almost any time of day. This seemed to me to be a real advance in fly fishing." As Gingrich says, when Hewitt came up with a good idea, he had the irritating habit of deifying his discoveries. Referring to Hewitt's claim for his skaters: "But really, doesn't he make it sound just a little more important than man's discovery of the means of making fire or the invention of the wheel?" On the other hand, the best objective documentation of the skater's prowess comes from Gingrich: "My old Cahill Bivisible Spider [skaters were also called "spiders"], the most versatile fly I ever found for stream fishing, was a near twin of Mr. Hewitt's Neversink Skater. I used to get them from Mills three or four dozen at a time, and once used nothing else for an entire season" (*The Fishing in Print*). (Spiders were identical to Skaters except for the way the hackles were tilted.)

To Hewitt's credit, he recognized that the skater was not for everyone or every fishery. They were difficult to cast and tiring to manipulate. And only large trout could get the two-inch flies in their maws. He viewed the Skater as a special-use fly. John Atherton, who had watched the evolution of the spider, was most laudatory. "If I had to be limited to one dry fly it would be the spider, without any doubt," he says, and "the spider is a great boon to the mediocre caster as it parachutes to the surface." He notes that drag, the fly on tippy toes, can be a good thing (*The Fish and the Fly*). Leonard Wright suggested that the trout might take skaters as crane flies or fluttering stoneflies (*Flutter, Skitter and Skim*, 2001).

CHAPTER EIGHTEEN

ENTER THE BIVISIBLE

Given his interest in biology, it's surprising that hatch matching was not Hewitt's thing with the exception of size. "It is important to note the size of the insects to which the trout are rising and to fish a fly of the same size or smaller. Trout generally care more about the size than the particular design of the fly and almost never refuse a fly because it is too small." Hewitt usually fished a #16; sometimes an #18. "Don't get a raft of patterns. They are not necessary at all and only confuse the fisherman into thinking he must have a certain pattern instead of studying what the general shade and size should be used to get his fish." Hewitt fished a small number of standard patterns such as the Quill Gordon, Cahill, and Whirling Blue Dun when hatches were on and liked to trim the hackle into a "V" on the underside, finding that this treatment led them to "cock" better. Otherwise he thought that a pattern with impressionistic appeal that would float well and was easy to see would usually do the trick.

Hewitt's famous Bivisible fit those qualifications. It was a simple wingless and tailless "hackle fly" not unlike older English patterns. Brown, palmered hackle covering most of the shank was relieved by a thin ring of white hackle just behind the eye. The trout were attracted to the brown body, while the angler could easily follow the white "neck." Other combinations featured a badger, gray, or black hackle, but Hewitt's original Bivisible was brown and white.

Was he proud of his creation? "This fly in various sizes is certainly the most universally useful fly we have and is perhaps fished now more than any other dry fly. The fly is by far the best of any I have yet seen for all species of trout and it is based on a sound physical principle. There is no fly which will catch as many trout as the Bivisible fly properly handled." You can see why these pompous claims might cause some irritation, and Hewitt had to admit that the Bivisible failed to win a popularity contest when the Anglers' Club members were polled. He often fished the pattern as a "searcher," just as an Elk Hair Caddis or Parachute Adams is often used for prospecting today. Hewitt thought that the fly might pass for a caddis or any of a number of terrestrials.

Unfortunately my collection of old catalogues doesn't go back far enough to trace the declining popularity of the Spider-Skater and Bivisibles. Ray Bergman's *Trout* lists his favorite patterns in 1951.

Two Bivisibles and three Spiders are included. It would appear that they slowly began to fade from view when Schwiebert's *Matching the Hatch* (1955) and other works initiated a rush of interest in more imitative designs.

HEWITT'S NYMPHS

Hewitt held Mr. Skues in the highest esteem. He returned from England in 1925 with a whole collection of his nymphs. He also scoured London's shops, purchasing as many other patterns as he could find. "I brought them back to this country with the anticipation that I would have wonderful sport with them. However, I was doomed to disappointment, as English nymphs only rarely take many fish in this country. This appears to be because the type of underwater insects in most of our streams is quite different from that in the English streams, where vegetation is more common."

Hewitt, although never a genus and species man, was an active rock turner nonetheless and quite aware of the physical characteristics of the larvae he found in the Neversink. He consulted experts at Cornell who stated that "80% were found in faster water and were of the flat-bodied, clinging type." He and his friend John Alden Knight achieved a flattened configuration by soaking the dubbed body in lacquer and pinching it flat until dry. The slender abdomens tapered sharply to an expanded thorax with short wisps of "handlebar mustache" hackle legs. Innovatively, they were two-toned, dark on top and a lighter shade of yellow below. Hewitt was delighted when a colleague fished a flat-bodied against a round-bodied copy, the flat nymph winning, 165 to 35. (It's surprising that he never spoke of mathematical models for determining statistical probabilities.) In accordance with the minimalist view, three patterns were deemed sufficient. (Knight rather than Hewitt may have been the inventor of the flat-bodied nymph, "having tied the first three patterns of this variation at Hewitt's Camp in the Catskills in the summer of 1931," Sturgis, *Fly-Tying*, 1940.) Ray Bergman's *Trout* includes color plates of Hewitt's nymphs in the 1964 printing, but they were disappearing rapidly from catalogues. One criticism was their tendency to flip-flop, top to bottom, and they were hard to cast.

As Schwiebert explains, Hewitt's stream improvements also led to warming water in the pools where swimming, round-bodied larvae began to take up residence, thus creating a different opportunity (*Nymphs*, 1973). Copies were tied based upon stomach-content specimens. We can presume that the pattern type was matched to the water being fished; however Hewitt often mentions his yellow-bodied "Stonefly Nymph" as a favorite general pattern, fished on a long leader in size 16. "When I want fish I always put this on." Eventually Hewitt stated that his nymphs were so deadly that a given stream could essentially be emptied with them. Without wholeheartedly subscribing to this grand claim, Gingrich confirms that his patterns were indeed effective. (Hewitt published the thin *Nymph Fishing For Trout* in 1934; I've not been able to secure a copy.)

WHEN IS A WET FLY A NYMPH?

Beginning with Stewart, that question has come up over and again. Hewitt had his own answer. His definition of an artificial nymph was rigid. It was a fly tied specifically as an imitation of a natural larva, whereas a wet fly, accepted by trout feeding upon nymphs, was still a wet fly, even if manicured to better suggest the naturals. John Atherton says there were evenings during the hatch on the flats when the big browns were oblivious to the seductive skaters. The ever-resourceful Hewitt then changed to a small wet fly, very thin and with the hackle clipped off top and bottom and a slim, sloping wing. He'd drop the fly ahead of a cruising fish under the bushes along the far bank "almost like leading a bird by swinging the gun ahead of him" (*The Fish and the Fly*, 1951). Hewitt wrote, "I often find that I have to cut away some of the wings to get the most fish— in other words, making the wet fly look much more like a nymph. This is really nymph fly fishing with a fly which is not altogether like any nymph in the stream. However if the fly is cut right it seems to work about as well as any nymph I have." In either case Hewitt stubbornly insisted that "nymphs" should be fished using the same tackle as with dry flies, fine and far off, but across and downstream, noting that drifting nymphs face into the current whereas, fished upstream, the imitation would drift "backwards."

Pools naturally required long casts, the fly several inches deep and drawn slowly with the last four feet of leader submerged. Under these conditions a trout will come twenty or thirty feet to the fly, all in the angler's view. Where the nymph could not be seen, Hewitt suggested attaching white wisps of silk to the leader as indicators while greasing the mid and upper part of the leader. In this instance the fly was allowed to sink two or three feet. When drawn, the nymph rose toward the surface—like a hatching natural.

> If they do not catch trout, it is because they are not fished right and do not move in the water in the natural way such insects move. In the swifter runs and especially at the heads of the pools, the fly should be allowed to float downstream, swing into the stiller water at the side of the current and then pulled up and down slowly a few times. Sometimes slight jerks given the fly during its passage around the swift current will produce an imitation of the spasmodic swimming of the natural nymph.

Riffles were fished with little or no slack in the line and sometimes seasoned with gentle rod-tip jigs before the final swing and lift. These presentations might equally apply to imitating caddis pupae, but Hewitt never commented. "I regard nymph fly fishing as a more skillful way of taking trout than the dry fly and one requiring infinitely more knowledge and technique. I can teach any one to be a fairly good dry fly fisherman in a few days but I would not undertake to make a good nymph fisherman in a year." Hewitt was convinced that smaller calibers were best and, convinced of the trout's ability to see in low light, didn't hesitate to tie on a #20 as darkness approached, when the nymph showed to the greatest advantage fished against the dry fly. He liked to fish wets and dries in competition.

> I tried for several seasons fishing a piece of water up with the dry fly and back down with a wet fly. One day I would go down first and the next day I would come upstream first so that each type of fly would have first chance at the water in turn. I found that for the whole season the wet fly would take about twice as many trout as the dry fly. Even when trout are rising freely to surface insects the wet fly will take them about as well as the dry fly if it is fished right.
> (A Trout and Salmon Fisherman for Seventy-five Years, 2004)

FISHING WITH LA BRANCHE

When it came to fishing the pockets and rapids, La Branche had no greater admirer. Hewitt repeatedly referred to him as *the* past master of the faster water, attributing much of his success to the "bounce cast that makes little explosions of light below the surface, and attracts the fish's attention." Hewitt practiced the induced rise on occasion, while his friend could flutter spiders with the best of them. Hewitt appears to have been fondest of the pools and flats, La Branche of the pockets and riffled runs, and in accordance, they preferred different tackle. La Branche fished closer and heavier; Hewitt further and lighter, even with the "switch" or roll cast. At one time both men touted fast, tip-action rods. Their preference later changed to a more moderate action, in part because the faster rod's tip took so much wear and tear. Their major disagreement may be reflected by La Branche's comment that "pools don't fulfill their promise." The men remained the best of friends, and since both fished for sport, I assume that Hewitt's declaration of the wet fly's supremacy was acceptable to his colleague. Indeed, La Branche admitted that the wet-fly method was more productive in the pools. The two often engaged in good-natured badinage, although on one occasion La Branche got a little rough, resulting in a temporary cooling of their relationship (*Catskill Rivers*). Their ideas about the color and wing structure in dry flies were similar and both recognized that natural-appearing movement of a fly was often the trigger that resulted in a take.

MIDGES? NOT A PROBLEM!

Hewitt had much more success with midge hatches. When trout are feeding on the surface and will not take any kind of dry fly, they are always feeding on midge larvae and midges. To him "midging" was simply a matter of fishing very small dry flies or nymphs. Commercial ties were too large and/or bulky, so he designed eight patterns to be tied on #20 and #22 hooks. "One of them will usually take any trout that's taking midges. Some were tied with hackle to float in the film. Others imitated the larval form, just below the surface, as the trout's back and tail, broke the film. I have frequently had great fun with my friends finding a bunch of trout rising to midge larvae." They would fish with every fly in the box and fail to get a single, definite rise. He'd then show them how midge-feeding trout can be "caught in any desired number." A hint: When you see a trout strike at a leader knot, "just put on a midge and have some real fun!" Hewitt did not anticipate that midging would become popular, since it was rather tedious and "because the tackle is too fine and few people have an unconscious feel of where their fly is located if they cannot see it. I myself, know within a few inches where the fly is in the water, and the fact that I cannot see it does not affect my fishing in the least. . . ." He claimed a "sixth sense" in setting the hook, really just "tightening."

TACKLE ADJUSTMENTS

Moving from a pool into faster water, leader changes were facilitated by winding the reserved leader around a hatband where bits of cork were embedded to hold the hook. (There have been some amusing consequences when I've tried this.) There was no one as fussy as Hewitt when it came to leaders. To assure that the taper in his self-designed leaders was accurate, he used a slotted gauge, for one of his axioms was that fish turn away from a fly for one of two reasons: either the fly was too large or the leader had been seen. Of course drag was another possibility, although as Gordon observed earlier, trout sometimes take a fly just as it begins to drag. Perhaps an insect struggling out of its shuck might minimally resist the current. Like La Branche, Hewitt practiced dropping the leader with a little slack and liked to put a curl into the very end as facilitated by a 5X or 6X tippet.

He fished "as fine" as he could for another reason. To him the ability to strike and play a fish on the lightest possible tackle was an important part of the sport. Even smaller fish could defeat the heavy-handed or careless angler using wispy tippets. His plan for setting the hook safely was derived from physics wherein rod length and weight, pull in ounces, and "safe strain" on the rod were derived as a function of the rod's angle. Conclusion: the rod should be nearly vertical where "the spring of the wood will cushion the blow to such an extent that very fine tackle will not be parted."

> The tip of the rod should be raised until the fish is felt on the fly, and then a very slight twitch given, which will set the hook and not injure the finest tackle. When I am in good practice, after considerable fishing, I am able to hook practically every fish that rises, and land him if I want to, in this way. It is easy to tell just how well one is striking because of the fact that a trout always turns when he has taken a fly, and he does not eject the fly at once from his mouth. Count every trout hooked in the corner of the mouth as a perfect strike, and every one hooked at other points as struck wrong.

About Salmon

"The art of dry fly fishing for salmon was worked out by myself alone in Newfoundland, and with my friends, Ambrose Monell and George La Branche, in New Brunswick." Monell was a capitalist in the metals industry, particularly nickel. (The Monell Foundation today supports a wide array of scientific projects.) Hewitt believed that the precepts involved in fishing for trout and salmon were much the same. He observed that salmon parr sip tiny insects, thus identifying a freshwater feeding habit that was confirmed later when smolt took insects from the surface. While salmon no longer feed after returning from the sea, Hewitt believed that salmon might be capable of continuing to act out this old memory. Why not try to trigger that impulse with a dry fly? (La Branche disagreed and refused to believe that salmon had much memory.) Hewitt nonetheless expected to hook 10–20 percent of salmon lying in the tail of a pool. Just as with trout, the way the cast was placed was critical, and the same concerns over leader visibility applied. His favorite pattern was a #10 Bivisible in gray or brown, and at times he employed La Branche's "induced hatch" technique. When salmon were rolling, winged patterns such as the Cahill were favored. He found no value whatever in attempting to imitate whatever might be on the water, believing instead that "light effects upon the fly above and below the surface" were the operative feature. Here John Atherton likely influenced Hewitt. In Atherton's *The Fish and the Fly* he applies the conceptual basis of the impressionistic school of painting—essentially the ways in which the reflection and absorption of light can provide an illusion of life—to pattern construction. Hewitt's description of "light effects" sounds very much like Atherton.

Hewitt may also have been the first to take salmon on nymph patterns after observing that they sometimes behaved like bulging trout. He agreed that the stomachs of adults are empty, yet the bulging suggested that larvae were being taken. Problem followed by hypothesis: Salmon crush larvae in their throats, spit out the carcasses, and swallow the nutritious juices! A professor from Bristol University verified this thesis by comparing fluids from salmon stomachs with "juices" released from crushed larvae. One bright, morning when there was little chance with a dry fly, Hewitt experimented, taking several fish on a nymph. That afternoon he and his son Abram continued to have great success with nymphs when standard fly patterns and methods were ignored. At times he fished a small wet fly tied to create a reflective flash or treated the fly with a little dressing to capture a shiny bubble, a trick we use today. (Hewitt once believed that the Neversink's cold waters could support salmon and stocked the headwaters; he was wrong.)

TACKLE: INVENTOR, INNOVATOR, *AND* ENTREPRENEUR

Hewitt would have been right at home working in research and development for a tackle company. With a representative of the Bakelite Corp he helped design the first versions of impregnated rods. (Bakelite was an early plastic.) He foresaw the coming of a synthetic material for fly rods and even attempted a composite blend of nylon with strips of wood and also small glass fibers stretched lengthwise along a soft wood core, the whole enclosed in strong plastic to form a hollow tube—this before fiberglass came along. Hewitt actually had his own tackle business, Trout Fishing Specialties, designed to advertise what he felt to be good yet affordable tackle and equipment.

Hewitt's Handbook of Fly Fishing mentions the following items:

Hewitt's Line Grease (containing talcum powder), 30 cents a box
Hewitt's Taper Line
Hewitt's Leader Soak, 50 cents for a 2 oz. bottle
Hewitt's Duplex; a combined insect repellent and dry-fly oil,
    50 cents for a 2 oz. bottle
A Hewitt fly box; no details
Hewitt's Jungle Fly; for dressing dry flies and as a repellent for
    black flies and midges, 50 cents for a 2 oz. bottle
Spider patterns, circa $2 per dozen (appeared in the 1936
    catalogue)

Mills also sold his popular Stonefly and Inchworm patterns. Hewitt even manufactured his own reels, but I'm not sure they were sold.

He argued against canvas creels (poor air circulation) and somehow knew that the enzyme pepsin in trout stomachs is capable of digesting the lining and is activated when the contents reach ninety degrees. Wicker creels were better but expensive, and he recommended the twenty-pound size, not for fish but to hold wet clothing and wading shoes. Better still, a light bag of woven grass, easily worn out, but easily replaced. And if only a fish or two was to be kept for supper, a red bandanna in the pocket sufficed. Hewitt claimed to have been the first to wear felt soles with wrought-iron hob nails way back in 1886. He owned many Leonard rods, each with a slightly different action and liked four-piece rods that could be transported in a "dress suitcase." As a consultant he argued for larger line guides, closer spaced. Instead of touting the top of the line ($55) he suggested rods in the medium price range. In the late 1940s he still preferred gut to the new nylon leaders, which were too stretchy and lacking in stiffness.

Catch and Release: First Glimpses

Hewitt claimed to be the first to fish flies in the Yellowstone River if not in the park. His father wanted to visit Yellowstone before it was opened to the public and arranged a private car on the Northern Pacific accompanied by Edward, then in his late teens, and the Secretary of the Interior. The tracks stopped at Billings, where General Phil Sheridan, fearing for the travelers' safety, sent thirty cavalry and supplies along. Hewitt took it upon himself to feed the whole group with cutthroat in the two-to-four-pound range. He depended upon live grasshoppers along the big river, later switching to flies. Traveling out of the Park to rejoin the railroad, Hewitt was asked to supply more fish for a military detachment, this on Boulder Creek, a tributary of the Madison that I've fished. He caught roughly 500 pounds of dressed fish in a single day. (I was pleased to release a dozen or so.) Further west along the Snake an encampment of Indians welcomed Hewitt's offer to slaughter fish for their winter's larder. (The men found fishing a chore.)

It's interesting that when Hewitt revisited the Rockies in 1914 the fish ran much smaller. Now fishing dry flies, he found that longer casts and delicate tackle had become necessary. In *A Trout and Salmon Fisherman for Seventy-Five Years*, Hewitt admitted: "I have no doubt killed more fish than I should at times." But those were different times, and I'd argue that Hewitt deserves recognition as an early voice for conservation in that his later successes are reported in terms of the large numbers of fish released. "The skill of playing the game is the real interest and the real sport. If more fishermen would play the game for the sport and return the fish to the stream, we would all have better fishing." Then, using real data, he attempted to dispel the notion that mortality rates run high when fish are carefully released. Some 700 trout caught by Hewitt in the Neversink during a season were transferred to his hatchery, their fins marked with a small punch. Remarkably, only two died. That's a good deal better than we do today, fishing barbless. "There is only one answer to making better trout fishing for everyone and that is to reduce the number of fish each fisherman takes per day and season. As trout caught can be returned to the water with very small loss of fish, the fishermen can catch their fish, return them to the water, and have them to catch another day."

After the market crash, finding himself "short," Hewitt began to rent rods, $125 for the season. Not surprisingly the takers included many of the best Catskill anglers. It wasn't a formal club, yet the assemblage became something of a round table. Alfred Miller ("Sparse Grey Hackle") was one of the knights. Hewitt was so protective of the large trout that came from his hatchery that when a paying guest hooked one, he'd wade out and release it, much to the dismay of the excited angler. In the 1940s the New York State limit was ten per day. Hewitt argued that a limit of four would be better.

THE ENGLISH CONNECTION

In 1925 Hewitt was invited, I believe by the historian Major John Waller Hills, to visit the famed Hampshire chalk streams. Hewitt privately questioned whether the fishing would really be as demanding as Halford and Skues had said. Afterwards he wrote, "The knowledge of the very great difficulty of the sport, and the fact that these trout can only be taken with the finest tackle and most skillful casting, makes this fishing more interesting than anything we have." The trip began with two days on the Itchen as a guest of Mr. Skues. Hewitt wanted to fish alone to better test his skills, free of advice. They were humbling days. Some actively feeding fish, affronted by Hewitt's well-presented dry fly, were put down. Of those he hooked, each demand a different pattern such that he never succeeded in raising two fish to the same fly.

Major Hills next introduced him to Halford's historic Houghton Club on the Test, where things were a bit easier although both he and Hills failed to score well. (Hewitt's spellings "Ichen" and "Teste" are unexplained.) Hewitt commented that feeding fish, holding in position close to the surface, were likely taking "larvae or nymphs" for the most part, whereas he and Hills persisted in fishing dry. The river keeper offered a plausible theory regarding the difficult fishing. The limit for the Houghton Club was twenty-four inches, smaller fish being returned. He suggested that the "easiest" trout were caught out each season, leaving the large, shy fish to serve as brood stock. And they bred well. Fed on mussels from the nearby sea, a sixteen-inch trout averaged three pounds! Later, when Hewitt was again a guest, he stuck with nymphs and caught many more trout than on the other visits. "When I want to catch trout in any quantity, I always use a nymph." Hewitt referred to the Club as the "finest sporting place I have ever visited" and remarked upon "a variety of fishing we can never have in America because we have no streams with these natural conditions." This comment later drew fire from anglers familiar with Pennsylvania's Cumberland valley streams and others in the West.

Legacies?

Hewitt showed that while a gentleman might prefer dry flies, he'd catch a lot more fish below the surface, whether he used flies or bait. In fact he stated that his flies would have no chance against a worm in the low waters of late season, fished upstream with a long rod as practiced in Stewart's and Pritt's time. Hewitt gave us permission to fish any way we wish, yet at the same time he was a major factor in the development of "high tech" small flies and long leaders with tender tippets. It's what you see today on our tailwaters and spring creeks.

While Hewitt didn't invent fiberglass rods he experimented with prototype composite materials and the concept of the hollow blank. And his preference for stocking fry foreshadowed the streambed nursery or the Whitlock Velbert Box of the 1970s. Hewitt blurred the distinction between wet fly and nymph by trimming wet-fly patterns into configurations more like larvae and fishing them in the same way as his nymphs. He acknowledged that the upstream wet-fly fishermen had reached the pinnacle, yet his downstream fine-and-far-off technique must have been far more demanding. Perhaps that was the attraction. It can be argued that Hewitt's dismissal of hatch matching impeded progress, and yet there was a place for an attractor pattern designed for "prospecting" between hatches as proven by his Bivisible. (I made it a point to fish Bivisibles off and on during a couple of seasons recently. They are truly productive freestone searchers, durable and fine floaters.)

Hewitt grew a bit testy with advancing years. In an argument over leader design with George Harvey, the renowned professor of fly fishing at Penn State, he said acerbically, "Young man, when you are as old as I am, you will realize that you were wrong" (*Catskill Rivers*).

A Trout and Salmon Fisherman for Seventy-five Years concludes: "It is wonderful to watch the evening light come on and the night finally take the place of day. This is what is happening to me; the evening is coming on pleasantly and the night will soon take its place according to the order of nature." He requested that his ashes be scattered in his river—"It will give the trout a chance to get even."

Preston Jennings,
America's Ronalds

*D*espite Theodore Gordon's plea for an American angler's entomology something like Ronalds's, our pioneers did little to initiate or support such an effort. Gordon would have been well qualified and was in an ideal position to begin the task. La Branche was a nonbeliever in imitation. Hewitt's spiders and bivisibles, while not flashy, qualified as attractors. But by the 1930s a generation of true biologists had taken to the water, university types devoted to pristine science and taxonomy, unsullied by the fly fisherman's more visceral and ulterior motives.

It was with this foundation that a team of Catskill anglers headed by Preston Jennings and, later, Art Flick produced two illustrated books that blended the fishing, the insects, and the science: Jennings's *A Book of Trout Flies* (1935) and Flick's *Streamside Guide* (1942). *Team* is an apt description in that the considerable time and effort that went into the project was carried out by a group of approximately nine like-minded colleagues, backed up by professional entomologists. The study was carried out over a specified period and on selected rivers. The experiment was planned not unlike a modern a research project (although without a budget). But before going on, let's wade up a short, historically related tributary and the story of an interesting contemporary angler.

THE ENIGMATIC LOUIS RHEAD

According to Paul Schullery, "Louis Rhead was one of the most creative, fresh thinking and stimulating American fly-fishing writers" (*American Flyfishing*). An Englishman, Rhead emigrated to Brooklyn in 1883 and earned a reputation as one of the finest book and magazine illustrators of the time. An avid Catskill angler, he illustrated many fly-fishing scenes. As an entomologist he spent seven years studying the insect life of the Beaverkill, some ninety-five species in all, each recorded in his paintings (*Catskill Rivers*). Other food forms were also of interest, and Schullery reproduces line drawings of the construction of a realistic shrimp imitation tied much as today. Rhead, who also fully understood that trout feed first upon larvae, then upon the adults, offered instructions for presenting nymphs, anticipating Hewitt. *American Fly Fishing* includes his diagram for imitating the upward rise of hatching larvae from the streambed. Published in 1916, Rhead's *American Trout Stream Insects* featured his series of American Nature Flies.

When it came to imitation, Rhead preferred to create an empirical artist's image, discarding the "science" as unnecessary. Rhead's entomology was apparently pretty superficial; for example, mayflies were lumped together as *Ephemeroptera*. More critically, his well deserved kudos as an illustrator led him to the notion that those same artistic skills might be applied to designing flies with equal success. Schullery reproduces a page from the William Mills catalogue wherein about a dozen flies each for the months of April though July are touted, supported by line drawings. The reader is tempted too by the ingenious Humpback Nymph tied with loop wing and a "reversed dry fly" to be fished downstream. All of these creations were patented. Emboldened by a superabundance of ego, Rhead also had an excess of entrepreneurial zeal, insisting that only select tyers could produce his line and that they could be sold only though select outlets. Worst of all, Rhead labeled his own imitations, ignoring time-honored, familiar patterns, and kept the recipes a secret. Anglers could hardly relate to labels such as Brown Buzz or Flathead. In short, Rhead oversold his products, meanwhile ignoring the beloved and dependable old patterns. Fatally, in the end Rhead took to criticizing the Catskill regulars.

(Jennings credits Rhead as the first to address American stream insects in book form.)

Jennings's Team

It has proven difficult to learn much about Jennings. We know that he was studious, considered a career in medicine at one time, and had great appreciation for art. We can guess that he had a talent for leadership and organization. I give Jennings and his team a lot of credit. There are lots of dedicated anglers, appreciative of the insect life relevant to their sport, who still aren't about to sacrifice their fishing time to turn over rocks or catch bugs in nets.

Their investigations took place in the Catskills, Adirondacks, and Poconos during the seasons 1933–35. The rivers mentioned most prominently include the Ausable, Beaverkill, Brodhead, Esopus, and Scoharie. Findings were credentialed by card-carrying professionals such as the respected entomologists James G. Needham and Herman Spieth. Unlike Halford, Jennings never presumed to invent a better mousetrap for each of the important naturals. Instead, he followed Ronalds, matching taxonomy to long-proven patterns.

A Book of Trout Flies was a timid title for an immensely useful work. The watercolor reproductions, while not as crisp as Ronalds's illustrations, were nonetheless praised by Schwiebert (*Trout*, 1978) and an eager and appreciative audience awaited. Schwiebert wrote the introduction to the 1970 edition, crediting Jennings as his inspiration for *Matching the Hatch* (1955) and noted, quite reasonably, the impossibility of including the insects of fisheries to the west. (Needham published *The Rocky Mountain Species of the Mayfly Genus Ephemerella* in 1927, so classification efforts were already under way across the country.)

The Mayfly (Dry) Still Sits the Throne

The title *A Book of Trout Flies* suggests a catalogue of a considerable array of patterns matched to various insects, something like the concluding part of Ronalds's *The Fly-Fisher's Entomology*. Understanding what his readers would primarily be looking for, Jennings featured the mayfly and quite practically, concentrated upon those "species" that produced a dependable, heavy hatch of long duration. The Hendrickson, Pale Evening Dun, Quill Gordon, American March Brown, Leadwing Coachman, and the American Green Drake were so honored. And rather than suggesting an assemblage of essential imitations after the manner of Halford's "Thirty-three," Jennings said that relatively few patterns should do the job. For example, Roy Steenrod's Henrickson was "matched" to *Ephemeraella invaria* and closely related species. While detailed dressings for some insects such as the Coffin Fly are given, considerable value is found in impressionistic patterns such as the Quill Gordon, Hendrickson, Hewitt's Spiders, and the related variants. With the exception of two figures on the cover, naturals and imitations are shown separately. There are just two additional color plates of patterns and three of mayflies. (The book begins with an overview of the relevant insect biology and concludes with a segment concerning tying, tools, and materials.)

Of Caddis, Stoneflies, Midges, and Minnows

The Grannom was mentioned in particular as an important early season source of food. However,

> . . . the writer is of the opinion that the great bulk of these flies consumed by the fish are taken underwater, possibly during the night, as he cannot recall ever seeing a trout definitely rising for the winged fly. In general the habits of the Stone-flies are not such that artificial dry-flies suggestive of them can be fished with much hope of success and the writer has limited his artificials to two, both tied as wet flies. Most of the larger species have the disagreeable habit of crawling out during the early hours of the morning, when the fly-fisher is usually in his humble and to be hoped, virtuous bed.

(*Virtuous* has a Waltonion ring quite unlike anything in the rest of the book.) Another strike against stonefly larvae: they consume quantities of the far more valued mayfly larvae.

The Black Gnat earns a two-page chapter in view of heavy hatches and rises in pools. A simple badger hackle pattern, hackle only, was suggested rather than a more realistic imitation. Jennings said that since the Catskill streams were "not especially productive of fly life," the minnow was an important food form, the Black-Nose Dace in particular. Polar bear hair was highly touted for its "beautiful, glossy sheen" although "difficult to obtain," and marabou wings for streamers were beginning to attract attention.

BELOW THE SURFACE

Jennings understood that the dry-fly man's sport necessarily follows the ascendence and emergence of (mayfly) larvae. "Were it not for this trip from the bottom of the stream to the surface, it is doubtful if we would ever have dry-fly fishing." He was conversant with the four mayfly nymph types (burrowers, clamberers, swimmers, and crawlers) and correlated each with a respective genus. However, he saw the pre-emergent, surface-seeking larvae as something of an aperitif.

> A trout feeding on Nymphs in the lower levels of water, or a fish grubbing around on the bottom of the stream, is a better prospect for the bait-fisher than for the fly-fisher; but it is when the Nymphs start to ascend to the surface to emerge as flies when the trout leaves his grubbing and starts to look for surface food. Trout have more protection from their enemies when they are deep in the water, but the eagerness with which they feed on the ascending nymphs causes them to ascend to the upper levels of the water where they take the newly hatched duns from the surface.

Jennings's point, the trout, now with whetted appetites, relish the duns above all else. He concludes, "If a fish can be taken on the dry-fly most anglers of the writer's acquaintance would prefer taking it by that method. If it could be predicted with a reasonable degree of accuracy just when a given species of fly was ready to emerge, an artificial copy of that specific Nymph might be useful; but so far as the writer is aware the first warning of the hatch is the actual appearance of the duns. . . . it hardly seems worthwhile trying to imitate any specific Nymph." Here he uses Hewitt's definition of a nymph imitation; he was aware that Hewitt had experienced disappointment using Skues's patterns. Experiments with the flat-bodied nymphs only began in 1931, and Hewitt's pamphlet *Nymph Fly Fishing* came out in 1934, so "nymphing" was still pretty much in the wings.

At the same time Jennings had nothing against the upstream wet fly and said about the Hare's poll: "alone it is also an excellent material for the bodies of Nymphs or wet flies which suggest Nymphs." "For upstream fishing, a trimmed Hare's Ear wet-fly with only the stubs of the wings left on, is about as good a Nymph as has come to the attention of the writer." (The idea that a wet fly such as the Hare's Ear might pass as a "static larva" had earlier occurred to Halford.) A color plate of fifteen wet flies displays the Light Cahill as an imitation of the Green Drake larvae. A Quill Gordon was his choice to suggest the Hendrickson nymph, and a large Leadwing Coachman fished along the edge of the stream was chosen when the Mahogany Dun was emerging. Jennings sometimes, however, fished a particular wet fly just because it worked. He stated that his own March Brown "makes a good general pattern of wet fly for use in riffles or when the water is roiled."

The Variant, an Interesting Pattern Type

Dr. William Baigent, a Yorkshire angler, may have influenced Hewitt in developing the Skater/Spider—or the other way around. Baigent's long-hackled, sparsely dressed Variants had a lot in common with Hewitt's creations. Baigent was one of the first hackle breeders, crossing Old English Gamecocks with Andalusians in search of the elusive "blue dun" that Gordon so coveted. His work went on for some forty years to be taken up by others after World War II (*The Fly*). As part of the project, some very long, stiff-fibered hackle was produced. A small, more standard hook was featured instead of the Spider's short-shank hook. Hackle fibers created a stabilizing tail. The shank was covered with silver ribbing, adding flash.

No doubt Hewitt would have known about the idea, for La Branche corresponded with Baigent and became very fond of his Variants. Jennings also mentioned the Variant as a possibility when Drakes were fluttering about pools and for fishing flats as Hewitt described with his Skaters. The concept is familiar; the long hackle suggested "wings fluttering alongside the body." Jennings surprisingly found: "Very often a large fly of the Variant type will prove successful when fish are feeding on these tiny flies [midges]." Dark Blue, Grey Fox, Cream, and other variously named Variants became quite popular in this country as displayed in catalogues on into the 1950s. According to Baigent, as sold by Hardy's, they also earned a following in New Zealand and France. In concert with Hewitt's Spiders, I had quite an affair with Variants back then. You'd get all kinds of fish coming up for a look—but not so many took the hook.

A Curiosity Settled

Norris, Gordon, Hewitt, and Jennings were certain that the insects of interest to anglers in England were sufficiently different from their American counterparts to render the British imitations of little value in our waters. As an example, Jennnings notes that the Alderfly copy sold here was much like the British and performed poorly as might have been expected given that the naturals differed in appearance. And when Reuben Cross, prominent Catskill tyer, challenged the trout across the water with our favorite patterns, they failed to impress.

On the other hand, mayflies are very ancient creatures that roamed during the Paleozoic period before separation of the primordial land-masses began to form the Atlantic. So theoretically the original brood stock should have been the same. Some 300 million years would have been plenty of time for genetic drift not to mention environmental effects. Jennings pointed out that the "Olive Dun," the most important of all English mayflies, was "about the scarsest fly in America." Still, some resemblances were identified. The Green Drake managed to hold onto the same genus, *Ephemerella*, as did the *Grannom*—Brachycentrus. There were enough differences in size or color, though, that Jennings attached the prefix *American* when he listed them, and the March Brown too. On the other hand he found that our Pale Evening Dun was nicely "imitated" by the Little Marryat, obviously born in Hampshire, while Marinaro called Halford's Male Black Gnat "one of the deadliest flies in existence." So certain English patterns had real significance here.

Soon After

Charles Wetzel published *Practical Fly Fishing* in 1943. Wetzel's emphasis was upon the streams of northern Pennsylvania, thus adding new species to the Jennings list. Schwiebert points out that the work might have gained a larger following if it had not been released during the war and had it not depended upon black-and-white illustrations (*Trout*, 1978).

Flick's Vade Mecum

Paul Schullery indicates that Jennings's book sold upwards of 80,000 copies, and so did Art Flick's handbook *Streamside Guide*, which followed in 1947 (*American Fly Fishing*). My copy of *Art Flick's New Streamside Guide* was published by Crown in 1969 and measures four by eight inches. At the core are twenty side-by-side color photos of a mayfly dun and its corresponding imitation, backed up in the text with black-and-white photos, information regarding hatches, and recipes for imitations. The treatment was similar to that in Jennings's book.

Flick's parents owned a hotel on the West Kill, a branch of the Schoharie River, where Art became manager in 1934. Flick's personal qualities made him a popular innkeeper, and in addition he became an outstanding fly fisherman and one of the half-dozen most talented fly tyers of the time. On this basis Schwiebert says that the West Kill Tavern became a sort of upscale anglers' salon and one of the most famous fly-fishing hostelries of all time (*Trout*, 1978). In addition, the countryside offered exceptional grouse hunting. The guest list was a sort of Who's Who and obviously included Jennings. It was Jennings who first interested Flick in entomology. Ray Camp, an outdoor writer for the New York Times first encouraged and then cajoled Flick into beginning his project. Once it was under way, Schwiebert says, Flick pretty much gave up fishing for collecting, his creel full of vials and fixatives.

The Streamside Guide covers the same duns as did Jennings plus a couple of others. The side-by-side display, a great idea, was nonetheless somewhat disappointing. The naturals were long kaput, and only occasionally resembled the artificial. On the plus side, Flick modified earlier imitations or invented his own, coming up with some real winners. His Red Quill remains a favorite; also the Black-Nose Dace. Despite his knowledge of the great diversity of insect life, Flick didn't like to fish a lot of patterns, depending heavily upon versions of the Variant, Cream or Dun, and the Grey Fox. If Jennings's adoration of the mayfly seemed excessive, Flick went him one better. Caddis simply didn't count, stoneflies got four, tiny pages, midges were ignored, and terrestrials barely drew a nod. Still, and certainly to his credit, Flick devised matching patterns for six important larvae including the March Brown, Hendrickson, Pale Sulphur, and Blue-Winged Olive. His *Steamside Guide* was also handier than Jennings's book, both in size and outline.

Schullery indicates that unfortunately, Jennings became jealous of his former pupil's success, feeling that Flick had copied him (*American Fly Fishing*). When you read the two books in sequence, the section on nymphs in the *Streamside Guide* jumps out as a substantial advance in concept and practice. Jennings saw little chance of matching the "pre-hatch" whereas Flick did just that and with nymphs tied to suggest the naturals. If the relationship had become strained at the end, the two men nonetheless set the stage for today's sophisticated and beautifully (digitally) illustrated angler entomologies.

Ray Bergman's Bible

*I*s there an angler's bible? *Trout*, published in 1938, came pretty close. Bergman's book went through four editions, the last, sixty-two years after the first, reaching six-digit sales. Schwiebert called it "the best comprehensive book on American trout fishing" (*Trout*, 1978). He remarked upon Ray Bergman's "warm, personal narrative style." The combination was a natural formula for success.

My copy, the ninth printing, came at Christmas in 1945. We lived in Denver, and I remember that season well, for most of my classmates, normal kids, spent the holidays skiing. Nerd-like, I stayed home with my new book. As Schwiebert suggested, to read Bergman was to get to know him in person. It reached a point where I thought of Bergman as a friend, as "Ray." That summer when my father caught me siphoning gas from our car, I explained, "Ray says shaved paraffin dissolved in gasoline makes a great fly dressing." I got away with a lecture on the consequences of aspiration pneumonia. At the time, before sprawl, Denver was a young city and growing fast. Still, there were no fly shops or fishing clubs. There were no books for beginners, although I devoured Bergman's column in *Outdoor Life* and Ted Trueblood's in *Field and Stream* at the barber's. We bought most of our tackle at the local hardware store, and, as I've mentioned, the fishermen I knew used wet flies tied on snelled hooks. Indeed, the dry fly was suspected of being an affectation. Fly-fishing's westward progress seemingly slowed and eddied while crossing the Great Plains. Admittedly, Denver was home to the successful Wright & McGill tackle Company, and the fine Granger Rods were made there, but there's no doubt that the art was far more developed in the East, with its entrenched traditions. This is why I, and no doubt countless others, owed (still owe) so much to "Ray" for his gentle, yet comprehensive instruction.

In the Business

Born in Rockland County, New York, in 1891, Bergman was so devoted to the regional trout streams that he never bothered to finish high school and was later recognized for his atrocious spelling. It was an unusual background for an editor-to-be. In his early twenties Ray opened a fishing store in Nyack. However the years following the war were difficult for the sporting trades, and when bankruptcy struck, Bergman went to work for Mills and Sons in the big city. I have understood that Bergman traveled widely in the East as a salesman, mixing business with fishing pleasure. He learned enough about fish, fisheries, and fishermen to begin writing not long after going with Mills. In 1933 he wrote his first book, *Just Fishing*. A year later he began to tie commercially and opened what was to become a most successful mail-order business. *Trout*, appeared five years after *Just Fishing*, serving as an update, revision, and expansion.

Bergman's Travels

Bergman's experiences gave the reader that wonderful "you were there too" feeling. They were simple stories that carried a lesson and somehow stuck with you forever—in my case, sixty years. I was surprised to find that Hewitt, the Princeton man, and Bergman, high school dropout, had very similar basic narrative styles. Tales from the Catskills and Adirondacks were staples, of course. But much to the pleasure of westerners, during vacations Ray followed the sun all the way to the Pacific.

The Owens River in California posed a challenge solved by walking the fly downstream along the bank on a short line. In a most exciting chapter Bergman encountered steelhead on the Umpqua in Oregon. Colorado's Trapper's Lake introduced him to cutthroat, and in Wyoming he fished Grebe Lake for grayling. I could relate to Ray's frustration during a trip with his wife and friends to Yellowstone. Trout were rising steadily all along the sensuous currents of the Firehole, quite obvious from the road. However Old Faithful promised to perform in just a few minutes, and there were the Morning Glory Hole and Paint Pots to admire. They had only one car. You can guess the rest.

The Encampment River in southern Wyoming was another favorite for experimentation. Bergman cleaned up there by stripping a bucktail through a boil of rising fish during a "dry fly" hatch. I first heard of the Adams, then a relatively new pattern, in another of his Encampment River stories. It was a special trip when I was eighteen and had a chance to fish the very same "Mica Mine" canyon, of course with an Adams, and caught my first multipound trout. Ray had a lot to say about fishing lakes in Quebec and in the 1950s became interested in spin fishing, exploring the Norfork Dam fishery in Arkansas. He had something for everyone.

SOME FUNDAMENTALS

Here are a variety of basic truths (in no particular order) that I first learned from Ray, a random few that flyfishers still fish by.

- Let a wet fly/nymph rest on the bottom for a bit before retrieving. (I picked up the hand twist retrieve from Ray.)
- Shadows panic trout, especially in shallow water.
- Briskly flowing water with a choppy surface can be safely fished at closer range with a shorter, heavier leader than slow-moving slicks or pools.
- A dry fly tied onto the leader some distance above a nymph makes a useful indicator when a fish takes the nymph. It was not his standard practice, but Bergman certainly predicted the future!
- Dry fly or wet, it's smart to "fish out the cast."
- When one bank is easy to fish from and the other is not, try the bank less traveled.
- After a heavy thunderstorm a streamer is a good bet.

What is common knowledge now wasn't in 1945, at least not to a beginner. Most important, Ray taught me to tie. There wasn't anyone else. Twenty-one black-and-white illustrations of vise, thread, materials, and his fingers plus some text were enough. Ray used half-hitches in lieu of a bobbin. It all went with a Thompson Model A vise that came that same Christmas. I've never become conceited about my flies, but they do produce and I still tie Ray's way. It's embarrassing to back out when asked to participate in a tying symposium, but they'd laugh at a half-hitcher (no bobbin) using a sixty-year-old vise. Hundreds of dollars a pop, today's multitasking dynamic tools would leave me all thumbs. (There's a photo of Hewitt seated at his vise in *Catskill Rivers*. I'm sure it's a Thompson Model A!)

Dry Fly Preferences

Ray's widespread travel and contacts with other anglers were such that his lists of preferred patterns were probably representative in the 1930s and for some years thereafter. He was an eclectic, depending upon a small cadre of dry-fly patterns, some general, others more imitative, meaning a fly that resembled the insect and looked good on the water. The Adams, with its mixed grizzly and brown hackle, was a perfect example while the Blue Dun or Blue Quill and Light Cahill, Quill Gordon, and Henrickson's from earlier days were "suggestive" imitations. The generalist Bivisibles and attractor Variants and Spiders fit another category. Meanwhile he was well acquainted with western waters where big Wulff's, and the buggy Irresistible were becoming increasingly popular. So the list had geographical balance. The range of hook calibers matched to each of roughly twenty patterns is of interest. Numbers 10, 12, and 14 dominated. Relatively few were tied on hooks as small as #16s. The general fly-fishing public had yet to heed Gordon's and Hewitt's suggestions favoring smaller flies.

Bergman on Nymphs

Here's something of a surprise. Bergman and Jim Leisenring, America's best-known nymph fisherman, were fairly close in age and worked some of the same rivers, yet Bergman makes no mention of Leisenring. Instead: "I don't pretend to be a good nymph fisherman; in fact I usually feel a bit uncertain when using them. Personally I do not care to fish with a nymph steadily because it is tedious and tiring. Instead I use it as an extra ace for the purposes of getting another chance at a good fish." Bergman said that 60 percent of fish that look at but refuse a dry fly will respond to a nymph immediately afterwards and about half of those will take it, "sometimes the best fish of the day." Talk about ambivalence!

This difference between the two men, so similar in circumstance, says something about both. As we shall see, Leisenring focused upon difficult fish in difficult settings with surgical precision. Bergman, much more the generalist, represented most of the rest of us. The historical point too is that a clear distinction between nymph and wet fly was accepted by then, and Ray was an acknowledged expert with the latter. He regarded Hewitt's hard, flat-bodied nymphs as an interesting contribution, although their opacity was seen as a drawback. Bergman and a colleague experimented with ways of building translucent bodies that could be colored using a combination of plastic materials not further characterized. A series is shown in the 1952 edition of *Trout*. More like lures than flies, Bergman's and other molded nymph imitations never caught on regardless of a realistic if manufactured silhouette, translucency, and color shading. The naturalness of the stuff of bird and beast won out.

How Things Were

In the 1930s Bergman still had one foot planted in the past while the other inched into anything new and different in fly-fishing. *Trout* boasted illustrations of something over 300 flies painted in stark colors by Dr. Edgar Burke. There are fifteen plates filled with postage-stamp-sized reproductions of wet flies, dry flies, bucktails, nymphs, and even steelhead patterns, all matched to a recipe. About half the patterns also appear in Mary Orvis Marbury's *Favorite Flies and Their Histories* (1890). Leafing through, you're impressed by this monumental display of the creativity and inventiveness basic to the fly-fisher. But to what practical purpose? Gordon, La Branche, and Hewitt had all persuasively argued the wisdom of pattern parsimony, making Bergman's grand display seem regressive. I can only figure that Ray was a salesman and hoped the plate's splash of colors and wonderfully named creations would add appeal. We're always up for the next wooden nickel.

In his book he recommended two dozen patterns each of wet and dry flies. Among the former were traditional brook-trout flies such the Silver Doctor and Parmachenie Belle and the Greenwell's Glory and Wickham's Fancy, ancient English favorites. The contrasting dry-fly collection was liberally sprinkled with Bivisibles, Spiders, and Variants. There is historical significance too in Bergman's eager acceptance of the "new" spin casting gear and bright and shiny lures. The second edition of *Trout* (1964) offers two chapters on tackle and technique plus a color plate of lures. Since then we fly-rod types have pretty much separated ourselves from the spin fishers, but after World War II the Dardevils and their kin were pretty exciting.

DOES RAY BELONG?

Bergman had relatively little interest in picking up and carrying forward what had gone before. He was neither a fly-fishing pioneer nor an innovator in the manner of the earlier Americans like Gordon, La Branche, and Hewitt. They talked to a far smaller and more sophisticated audience. Nonetheless, Ray was observant and thoughtful, helping to entertain, instruct, and thereby recruit untold numbers of common folks to fishing, and not necessarily just for trout or even with the fly. *Just Fishing* sets that tone and, incidentally, went through thirteen printings, the last in 1949! Ray was respected by his talented peers, too. Charlie Fox of Pennsylvania fame bought most of his flies and other equipment from Ray and stated that *Just Fishing* had been his bible. "He came up with stuff we never heard of you know" (*Limestone Legends*, 1997). In that book Bergman struck a prescient note that has since swelled into a symphony of opportunity for some of us: "I could not help but thinking how wonderful it would be if all anglers' wives loved fishing. Certainly it would create a bond between them which none other could equal." That's pretty powerful, a bond, stronger than even our children! My take is more selfish. Let's say I want to fish for a month in New Zealand. If my wife does too, we go. If not—maybe not. Anyway, for a lot those of us between, say, forty and eighty, Bergman was a friend and tutor. He surely had a place in *our* fly-fishing history. ━━◁

Sawyer Elevates the Nymph

*L*ondoners, Halford and Skues and many of their confreres were fishermen first and foremost, naturalists second, and then primarily to the extent that insect life was central to their sport. Frank Sawyer's background was quite different. Born in 1906 on the banks of the Avon, another of the celebrated southern chalk streams in Wiltshire, he spent his life with his family on the river. One might say that the Avon, its banks, and the inhabitants thereof *were* both life and family. *Keeper of the Stream* (1952) reveals Sawyer to be as much a naturalist as a fisherman. His formal schooling ended when he was thirteen. He worked for a time as a farm laborer and later as an assistant river keeper until in 1928 he was precociously promoted to "head keeper" on the Officers' Association Water, a stretch of six and a half miles of the Upper Avon.

Sawyer had two passions, a great curiosity about and love for nature coupled with the desire to set down his feelings and discoveries in words. It was a frustration given his limited schooling until in 1932 he met an angler who was destined to greatly further his career. Sir Grimwald Mears, highly educated and socially prominent, respected Sawyer for his abilities on the river, and when he learned that the young keeper was struggling to write, Mears became a tutor of sorts. Realizing that Sawyer had true talent, he critiqued his work, and when nymph fishing arose as the topic of some of Sawyer's essays, he loaned him Skues's first books. Mears must have been surprised when the young keeper audaciously found several points of disagreement with the great man! Mears nonetheless shared the comments with Skues. Knowing that Sawyer's opinions were based upon real experience, Skues graciously responded: "Encourage this man Sawyer to write articles on the subject and I will arrange for publication." (Sawyer had the opportunity to meet Skues, then nearly ninety, at Nadderville, and they corresponded for several years thereafter.) Mears continued as Sawyer's patron saint and was influential in setting up his appearances on BBC radio fishing shows. Terry Lawton's *Nymph Fishing* (2005) is an excellent reference source.

The Nymph Violates Dry-Fly Water

At the time Halfordian purism was at its peak, and the nymph was generally proscribed on club or leased waters. During his first year Sawyer met Brigadier General H. E. Carey, another man who strongly influenced and promoted his career. Carey must have been a freethinker, for he had accepted Skues's wisdom and convinced Sawyer that the capture of challenging fish during the becalmed, bright days of midsummer was the epitome of the sport and that this required the adept presentation of the nymph. Sawyer was further encouraged to collect and copy larvae. Meanwhile Skues introduced Sawyer to Martin Mosely, who showed him how to classify nymphs.

Through his friendship with Carey, Sawyer was able to play a role in the revocation of the Dry Fly Only law within the Association. Sawyer was likely regarded as an inferior by some of the membership in that he was paid (very modestly) to work in their behalf and lacked the social background of the various majors and colonels. Early on he had to request the privilege to fish during his spare time. Fortunately for him, his mentor was a general! Besides, Sawyer's knowledge of the river and his skills as a fisherman demanded respect. In addition to his duties as a keeper, Sawyer served as something of a "guide in residence," instructing members when asked for help, whether with dry fly or nymph.

The opening of the Association waters to the nymph, or upstream wet fly, prefaced a gradual return to a more balanced view of fly-fishing's proprieties during the 1930s. Sawyer's goal was never the conversion of the dry-fly man, but rather achieving an understanding and acceptance of the nymph—in its proper time and place. Attacks upon Halford were not Sawyer's style, although in sly recrimination he suggested that there was no reason to scare the fish with futile dry-fly presentations when the nymph is "a better representation of the food the trout are looking for than is the floating fly."

Sawyer tied thinly dressed dry flies as well and enjoyed fishing them during Mayfly time in May and June. He often joined friends on the nearby Kennet for the event, commenting, "the dry fly can be very effective and no one would wish to use a nymph in preference." Later in the season, as hatches dwindled, he seems to have turned largely to the nymph, warning, "It is not an art to be acquired in a season or many seasons."

The Pheasant Tail Nymph

When you think about it, relatively few fly patterns can be confidently attributed to a particular tyer. Lee Wulff and his hair-winged namesakes, Al Troth and the Elk Hair Caddis, and especially Frank Sawyer and his Pheasant Tail Nymph are among the few. Uniquely, the pattern was fused at the hip, with a mode of presentation that was rarely seen at the time. Capriciously, Sawyer's definition of *nymph* was restricted to mayfly larvae although he fully understood that trout and grayling feed on all sorts of other larvae and shrimp. The mayflies were classified, à la Hewitt, into four types: swimming, crawling, burrowing, and flat-bodied clingers. Of these only the swimmers could be consistently imitated with Sawyer's method. The P.T., as he called it, was primarily tied to represent the swimming larvae of a group of mayflies traditionally known as Olives or Blue-winged Olives, most of them *Baetis* species (*Nymphs and the Trout*, 1958).

Like Skues, whose nymph fishing began with a sunken dry fly, Sawyer found that when "badly chewed," his favorite dry-fly pattern, the Pheasant Tail Red Spinner, was readily taken by fish feeding on larvae. Given that the tattered fly was a poor representation of the natural, Sawyer joined the impressionist school of fly design and offered a favorite axiom: "Many insects, few patterns." He wanted "representation satisfactory from the trout's point of view," believing that a well-conceived pattern should be simple in construction and serve to "imitate" a number of species, a dozen at least in the case of the P.T. "Simplicity is an aim to be desired." He concluded that the three essential features were form (tapered shape with a prominent thorax and wing-case bulge), color (reddish brown typical of the olive larvae), and size. After considerable experimentation he concluded that fibers from a pheasant's tail provided the desired color and were easy to work with. They were all he needed—almost.

BALLAST AND DURABILITY

The Pheasant tail nymph had a mission. Sawyer had observed trout taking motile larvae along the streambed and higher in the stream's mid zone. Simply, then, he wanted the P.T. to drop delicately and sink promptly, drifting deep with the current. At the time it was a novel strategy. Cotton said that trout are top-water fish. Stewart kept his spiders close to the surface. Skues's bulgers darted about not far below where Hewitt "drew" his wet flies to attract rising fish in the flats. With the exception of Hewitt's fuse-wired flies, the idea of the "deep" wet fly or nymph was still new. The Pheasant Tail was a good imitation for color, and its clean, minnow-like form was much like the naturals'.

There was a problem, though. *Baetis* larvae are small and, of necessity, so are their imitations. There was no room for fuse wire on a #16 or smaller hook, and Sawyer refused to attach shot to the leader. The solution was brilliant. Instead of thread, he tied the P.T. with fine "not much thicker than a human hair" red-colored copper wire found in small transformers or generators. The wire was durable and added ballast. After covering the shank, he tied in four pheasant "fibers," points protruding as tails. The remainder was spun with the wire and lapped up to the eye. A thorax/wing case lump was built up by folding the fibers back and then forward again to the eye—a simple, tight silhouette of the Mottram mold. (When more weight was desired, a lump of wire could be added under the wing case.) Because the larvae swam like tiny fish, legs swept back along the body, Sawyer saw no need for hackle and noted that the sleekness of his nymph, coupled with the added weight, allowed it to sink quickly. Meanwhile the feather edges protruding from the "twist" suggested the translucency imparted by a thin layer of air under a larva's loosening shuck. He fished the P.T. in three calibers, #1, #0, and #00, depending upon the water's depth and current, pretty much our #s 15, 16, and 18.

When I checked, the mail-order catalogues of four major companies all displayed tempting color photos of Pheasant Tail Nymphs, eleven versions in all, variously decorated with bells and whistles such as a bead, flash back, or rubber legs. A popular recipe is tied with thread, the thorax replaced with peacock herl with the butt ends of the pheasant tail fibers serving as a wing case or trailing legs. Are we to assume then that the original P.T. needed improvements? It's hard to imagine that American species are that different or that mutations over the years rendered the original P.T. obsolete. Obviously it's not. Tied just as Sawyer described, the fly is still a proud, productive, and upstanding addition to any fly box. It's we tyers who have mutated!

The Grey Goose was the P.T.'s primary running mate, lighter in color, more gold, but tied in the same manner. When one fails to attract, the other is often successful. But when no response to either is obtained, I consider myself beaten. Sawyer discusses and gives the exact recipe for both patterns in *Nymphs and the Trout* as well as in *Masters on the Nymph*, edited by Migel and Wright (1979). *Masters on the Nymph* similarly details Sawyer's other patterns. The plump-bodied "Killer Bug" was his regular means of decimating schools of grayling and proved productive in still waters. His friend the well-traveled Lee Wulff gave it that name. One more Sawyer pattern, the "Bowtie Buzzer," served as an imitation of large cranefly larvae, completing Sawyer's basic team.

Ritz and a Testimonial

Sawyer had attracted a good deal of attention through his articles in *The Field*, *Salmon and Trout Magazine*, and others well before his books were published. In 1952 Charles Ritz, scion of the hotel chain, developer of the parabolic rod, and world champion caster, dropped across the channel from his usual haunts on the French chalk streams for a visit. Instead of just chatting, Sawyer took Ritz fishing. The following comes from Ritz's *A Fly Fisher's Life*:

> When we came to a reach of calm water with a depth of approximately four and a half feet, I noticed rises, small rings, regular and spaced out. I was under the impression that they were rises after midges. Sawyer indicated that the fish were taking nymphs well beneath the surface. He cast upstream and placed his nymph about six feet from the selected ring. The nymph sank at once. Visibility was good and we could follow easily the floating part of the (greased)

leader which grew shorter during the drift. Sawyer explained to me that you must watch the leader at the point of immersion for that is where the smallest reaction will become manifest when a fish touches the nymph. Little if any slack was allowed to accumulate. Suddenly he very slightly raised the point of his rod, a movement followed immediately by a delicate strike, and he had taken it. Sawyer explained: When I thought that my nymph had passed the fish without being taken I slightly tightened my line *to give animation to the lure*, which often induces a fish to take.

Ritz was impressed that the trout were feeding too deep for him to see and yet still created small rings. His opinion was that the presentation involved coordination of four factors. The stream's depth and current speed were fixed factors. The weight of the nymph and hence rapidity of descent could be controlled, and also the "lead" or distance the fly is cast above a ring. We must assume that solving the equation came as the result of a great deal of experience for while Sawyer took fish after fish, Ritz struggled.

This description from Sawyer's *The Nymph and the Trout* is perfect.

One develops an awareness which is not even a sixth-sense. It is something which just can not be explained. You see nothing, feel nothing, yet something prompts you to lift your rod tip, some little whisper in your brain to tell you a fish is at the other end of your line. But this feeling only comes if you are intent upon your work, for though it may not be possible to see through the surface, it is possible to visualize the position of the fish and anticipate his reactions to your nymph.

Later Sawyer suggested that they fish by sight in shallower water. The cast had to be spot-on, without lining or alarming the fish. Ritz likely had no difficulty here, but because the fly couldn't be seen, the trout had to be watched for the white of an opened mouth or movements suggesting that the fly was being taken, and these were very subtle. Meanwhile the "set" had to be almost instantaneous, requiring lightning fast reflexes. Ritz had a wonderful afternoon:

I am eternally grateful to Sawyer who revealed his secrets to me with joy and enthusiasm. What I like about Sawyer is that he refuses to allow himself to be hypnotized by theories of over-precision. He is always in search of the simple and practical and the essential and attaches importance above all to the *method of presentation* (the type of leader, the weighting of the artificial nymph and the exploitation of the various possible stages in the development of the insects).

Ritz went on to Austria and Bavaria, trying his new skills for grayling and spreading the gospel. He later designed and named a rod after Sawyer. Thanks to Ritz, Sawyer became a member of the famous Fario Club in Paris and was privileged to fish many of the classic French chalk steams where the very best dry-fly men applauded his success with the nymph.

THE SINKING NYMPH

"The stomachs of rising trout are often choked with nymphs," according to one of Sawyer's favorite aphorisms. The point is repeatedly made that larvae, *Baetis* in this case, rise from the bed to the surface in an undulating, jerky manner, often to return temporarily to the streambed, and that they are frequently taken during their descent through the water column. This he observed during hatches when the trout were gorging on the duns, explaining, "The eyes of a rising trout though interested mostly on what is happening on the surface, are also keen enough to see movement beneath it." And this was the case in the shallower water he fished with Ritz. Long casts and more lightly ballasted flies were necessary, and Sawyer always fished a long, fine tippet regardless. Thus the Pheasant Tail served him at every level save the film itself. Politely critical of Skues, Sawyer observed that *Baetis* larvae drifting on or near the surface constantly struggle to free themselves from the shuck and attributed Skues's success to the movement of the hackles on his flies that simulated an insect's busy legs. In part he referred to insects with their setae stuck to the shuck as "cripples," as Swisher and Richard later called them. But this was not a stage for which the P.T. was well suited. "Seldom have I been successful in trying to interest trout with nymph patterns which float semi-submerged, just awash so to speak" (*Nymphs and the Trout*).

Sawyer would "fish the water" when necessary. "When trying to spot fish, read the water first to sort out the most likely holds (in this case current lines between weed beds), and only then concentrate on these select target areas. Look well into these places first and then look a second time." In the introduction to *Keeper of the Stream*, Sawyer's son Tim wrote, "My father saw the fish because he knew what he was looking for." He meant that before you can see a fish, the image has to be separated from its surroundings. It was a talent already highly developed when Sawyer first fished with General Carey and others who, lacking the ability, had more limited success with deep lying fish. It's fair to state that Sawyer thought that this ability to spot fish was more critical than any other (*Nymph Fishing*, 2005).

THE AVON ECOSYSTEM

I had been unable to find a copy of Sawyer's definitive *Keeper of the Stream* until in 2005 Tim and Nick Sawyer, Frank's son and grandson respectively, issued a limited edition of the original work, appropriately dedicated to Grimwood Mears. This charming book doesn't address the catching of trout directly. Rather Sawyer writes, sometimes introspectively, about his "classroom," the river and its complex ecosystem. With the exception of Gordon, I've seldom associated fishing writers with adjectives such as *caring, compassionate, kind,* or *sensitive.* That's no doubt unfair, but the focused expert usually sees the fishery and its trophies as a challenge to be attacked and overcome. Sawyer's charge was to protect, even nurture, the Upper Avon and its wildlife. "The river keeper never need be lonely," he wrote, "for he learns to understand and appreciate the wildlife which are his constant companions, and as he gets to see the fruits of his labors all around him he gets a certain satisfaction. In all Nature there is a reason for everything and in trying to assist her we must find out these reasons and try to follow them through." When a massive storm blew the Avon out of its banks, Sawyer hurried to assist bird and beast, the tiniest mouse, vole and shrew, the moorhen and chick to the safety of higher ground. There was true remorse when irrigation gates were mistakenly opened resulting is a sudden, short-lived burst that transported various creatures into the flooded water meadows, where many perished, and tiny trout became trapped in the depressions made by the hooves of cattle. Perhaps I am a sentimental fool at times.

Pike and grayling were another matter. Allowing pike to coexist with trout was likened to "raising chickens and foxes in the same run." Their murder was calculated and vengeful—caught with bait, speared, snared with a wire, and best, when clustered in shallows, blasted with a shotgun! Similarly, traps hung on wires across the river controlled the local populations of kingfishers, cranes, and heron.

Sawyer was an exceptional nature writer. At times his fishing seemed almost something of a spin-off. I highly recommend *Keeper of the River* to anyone, fishermen or otherwise, interested in quality outdoor writing. Without question, the book broadens one's appreciation of the man, and the color plates help the reader to visualize what the old water meadows were like.

Scattering Pearls upon the Water

Sawyer's wisdom comes through so many carefully recorded observations, bits and pieces that are scattered about and hard to fit into particular topical paragraphs. Many are found in *Nymphs and the Trout*, where the style reminds me of Dickens. The chapters, twenty-two in all, reproduced from pieces in *The Fishing Gazette* and other sporting periodicals, are brief, each supported by a cluster of provocatively worded subtitles such as "Frustration on the Test" and "An upstream tactic breaks the deadlock." The writing is fast-moving and enjoyable. The text never drags. Another reader would certainly have made different selections, but I offer some of my favorite Sawyer observations.

Concerning trout and other fish:

- Really big trout aren't hard to catch because they mostly feed on the bottom, losing their fear of anglers. When they do rise, it's with focused intent.
- Trout feeding in dense weed beds make a sucking sound when they strip the tendrils of vegetation through their teeth to scrape off clinging larvae. It pays to listen!
- Trout nearing the end of their lives and those that have been injured turn almost black, and become blind.
- When "over-played," trout die out of excessive fright, not fatigue.
- It's a mystery to me how trout are able to discriminate between men and cattle or dogs. Yet they can and do.
- The deep nymphing technique also worked well for salmon. Cast upstream Sawyer's grayling pattern, the Killer Bug, was typically taken just when he raised the rod tip to make the fly "deadly."

- Ever curious, Sawyer wondered if other fish would be interested in his nymphs. They were and on light tackle. He had nice things to say about sizable chubs: "They are no man's fool and truly 'not as daft' as they look." The Killer pattern did the trick. He had adventures with carp, lately targeted by some of our fly fishermen, and even roach, dace and perch. "Rough fish" provide a lot of sport in England today.

Concerning insects:

- Sawyer believed that nature had a grand plan that favored the trout. For instance, the role of the midge was to produce larvae as a source of food for the fish in the early spring when their egg sacs were depleted and other food sources had yet to develop.
- Mayfly larvae received more respect. He believed that they could sense and respond to danger. He had many times seen a trout ready to sip in a nymph drifting in the surface film, when at the last instant, the insect would molt and fly away to safety.
- On the other hand, Nature sometimes played nasty tricks. A heavy spinner fall came on after a rain, water collecting on a recently surfaced road. The lighting conditions turned the road into a shimmering, silvery band that the poor spinners misinterpreted as a stream, The roadway was littered with their bodies.

Regarding the artificial nymph:

- An artificial nymph is more readily taken and held by a trout if it has first been well covered and soaked with the slime. I've no idea to what extent Sawyer followed this practice.
- Downstream nymph fishing is never very profitable.
- During a heavy hatch on a windy day the trout let the duns go by in favor of the larvae. He assumed that the fish had become frustrated when the wind blew the duns off the surface so quickly.
- During a spinner fall the trout took nymphs instead. He observed that with their wings collapsed and plastered to their bodies, the outline was more that of a nymph.
- Eddies with a surface coat of scum should be fished with a nymph. The current-borne insects collect there and attempt to hatch, but fail because of the scum.

- When trout are concentrating on nymphs, it is difficult to interest them in a floating fly. However in the reverse situation, the nymph will often prove more productive than a dry fly.

Tactical:

- Sawyer preferred to face into the sun, trading off any protective blinding effect for the relief of concern over shadows cast by the line and leader.
- He was convinced that trout can see for considerable distances and would more often come for a fly from afar if it weren't for laziness—"too lazy to range very far." They preferred to wait for food to be delivered to them by the current. In *Nymphs and the Trout* it appears that he is referring more to nymphs than to dry flies.
- When looking for a subtle take, Sawyer greased the leader and for a visual "fix point" focused on the "little hole" the leader made as it penetrated the film.
- When setting the hook, a lateral flick of the rod was preferred to an upward sweep.

OLIVER KITE

Terry Lawton's recent book, *Nymph Fishing*, tells us a lot about another noteworthy English nymph fisherman. Oliver Kite, Sawyer's pupil between 1958 and 1961, was arguably as well known in his country and Europe as Sawyer, from whom he learned his techniques and tying method. Kite went on to write extensively and often spoke on radio and TV, where Lawton says he was a natural. Understandably the two men presented nymph fishing in much the same way and had similar talents.

I get the sense that Sawyer's rod tip elevation was quite subtle, whereas Kite is credited with introducing the more physical "induced take," borrowed from La Branche, for attracting a trout from some distance. As we know, that catchy phrase was made famous many years earlier by La Branche with regard to the dry fly. In practice the theory and technique were apparently much the same. For instance when unable to spot a fish in a likely-looking hold, Kite would attempt to induce the trout to move into a visible position by "animating" the fly upwards or to one side. And he furthered the importance of silhouette in nymph imitations with his "Bare Hook Fly": copper wire wound on to create a pronounced taper with a prominent wing case and otherwise devoid of other tying materials! The added weight was perfect for penetrating deep water. The two men differed in that Kite regarded large flies such as the Killer Bug as lures, lethal but unsporting. Otherwise, Kite and Sawyer were almost perfect parallels in their careers as guides, writers, and broadcasters, and both fished extensively in Europe and Scandinavia. Kite's only book, *Nymph Fishing in Practice*, was published in 1963.

For whatever reasons, Kite fell from favor with Sawyer and his wife Margaret in later years. Mrs. Sawyer felt that Kite had utilized what he had learned from Frank to gain fame and fortune. (Sawyer refused to participate in TV shows of the sort that showed Kite to such advantage, feeling that would demean the sport.) Adding insult to injury, Kite parked his ostentatious white Jaguar across the street from his neighbors, the Sawyers. Lawson concludes that "Kite was an interpreter of other people's, ideas, particularly Sawyer's, rather than an original thinker."

A Eulogy

Despite his fame, the humble river keeper and his wife lived a tranquil, pastoral life. He died on the bank of the Avon in 1980 while walking his dog, no doubt deep in the study of his river, as always. John Randolph had this to say in an editorial (*Fly Fisherman*, September 1973): "Among the princes of lineage and industry who sought his favor, he stood out as a man of great presence and quiet dignity, a person who without words captured an assemblage by his mere presence. He was a natural man among sophisticates, the teacher among teachers. . . . "

LEISENRING AND THE
AMERICAN NYMPH

*L*eisenring was known affectionately as "Big Jim" among the members of the Brodhead Fly Fishers Club. Schwiebert called the assemblage the Twelve Apostles. It was a fellowship of prominent anglers such as John Alden Knight, Preston Jennings, and Ed Zern that met regularly at the Rapids Hotel at Analomink on the Brodhead (*Nymphs*, 2007). Leisenring's role as leader was based upon his exceptional angling skills coupled with a kindly, humble nature. He was born in Allentown, PA, in 1878 and worked there as a tool maker, a proud, blue-collar trade of the Pennsylvania steel mills sort. Schwiebert reports that Big Jim was a large man, fond of his brautwurst and beer, and likely to go fishing "whenever the spirit moved him." If that meant losing a job, so be it. His skills as a machinist were so touted that another opening was always waiting. Like Stewart and Skues, he remained a bachelor. Equally adept with dry fly and wet, Jim had a special fondness for soft-hackled wet flies fished to imitate active, pre-hatch mayflies and caddis. His only publication, *The Art of Tying the Wet Fly*, was not widely recognized when it was published in 1941. The slim volume was primarily focused upon how Big Jim tied his delicate flies rather than his method of fishing them. In 1960 Vernon "Pete" Hidy, one of the younger apostles, encouraged *Sports Illustrated* to produce a commemorative article honoring Leisenring. Later Hidy polished the original manuscript and re-issued *The Art of Tying the Wet Fly* in 1971, adding his own second part, "*& Fishing the Flymph.*" (Leisenring had passed away twenty years earlier.)

His Wet Flies and Nymphs

There's a question of whether Big Jim is more remembered for his delicately crafted flies or the "Leisenring lift," an imitative method of presentation. Both reflect the precision you'd associate with a machinist working to close tolerances. As a tyer he paid close attention to qualities such as size, conformation, texture, flash, color, and translucency. The spun body is particularly associated with Leisenring—a dubbing of fur spun on silk thread into a tapered configuration. He often prepared sets of bodies stored on cellulose cards, waiting to be wound on. The instructions in his book on spinning bodies fill an entire chapter, including photographs, and are extremely detailed but hardly more so than the choice of the silk and wax (separate chapters) and the choices of fur selected from various animals. Schwiebert wrote, "The old fly fisher carefully noted the ingredients of each dubbing mix and chose its under-color silk to enhance the color of the dubbing in imitation of specific aquatic insects until it seemed to glow with an ethereal, life-like hue when wet" (*Nymphs*, 2007).

Leisenring preferred the thinly dressed British-style nymphs: "My experience in imitating and fishing artificial flies has shown that and proven to my own satisfaction, as well as many of the finest fly fishermen I know, that the wing is the least important part of the fly." As did Hewitt, he believed that wings could be simulated by hackle. At the same time, Leisenring paid a great deal of attention to feathers—to be used as hackle. There's a lengthy chapter devoted to the latter replete with a diagram of a bird broken up into labeled parts something like a beef in the butcher shop. There are even photos of the outside and inside of a wing, also with labels—primaries, secondaries, coverts, and the "bastard wing." He insisted that hackles have a reflective iridescence and the correct degree of stiffness for the water to be fished. Tyed in the soft-hackle mode not unlike Stewart's spiders, various feathers were chosen for their dynamic mimicry. "The careful fly tier will select his hackle according to its ability to *act* and *look* alive." He found no need to name his patterns individually, for each was designed in anticipation of a particular hatch and the creations were his own. As far as I know, he did not plan to sell them.

Leisenring's personable suggestions come to you quietly, as though from across his tying bench. Some are homespun. Here he writes about the fly tyer's nemesis, the moth: "While speaking of materials it is well to remember that furs and feathers should be placed in envelopes and cedar boxes and kept where sunlight or bright light may shine upon them. You should then have very little trouble with moths, because moths do nearly all of their dirty work in the dark. Remember that an evildoer loveth darkness (this also applies to very big trout)."

Choosing the best hooks was critical, and that meant a small barb. A "wickedly big" barb prevented the hook from penetrating beyond the barb or otherwise created an "enormous gash or hole when it enters," allowing the hook to slip out. Smaller caliber hooks were also preferred. Today we agree, if for a different reason. One's image of Big Jim belies the dainty, dubbed-body flies for which he became known. I couldn't help but remember an old ad for piston rings dubbed: "Tough but oh so gentle!" A fearsome-looking giant of a man was depicted tenderly holding a ring between thumb and finger.

The Wet Fly and the Nymph

If the Colorado anglers I knew back in the 1940s were asked what the trout take a wet fly to represent, the answers would have included a drowned bug, an immature aquatic insect and "Who knows? They just go for them." At the time the wet fly and nymph were generally discussed in separate chapters, witness Hewitt and Bergman. Leisenring followed suit and also accepted Hewitt's definition to the effect that a nymph should be tied with the specific intent of copying a particular "premature." Leisenring wrote: "Angling with the artificial nymph is one of the most interesting and effective methods of catching fish known to man, as well as the least understood in the minds of fishermen." Recipes for seven patterns were given, mostly mayfly larvae such as the March Brown. He used heavier hooks in the interest of reaching the streambed and had no use for a weighted nymph "because they do not swim naturally."

There was a distinction between streambed larvae and those near the surface. With regard to the former: "My experience has led me to believe that the trout is not so particular when he goes nymphing. He will often take any nymph he can dislodge, case and all." With reference to the subsurface: "At such times the fly fisherman is up against a different proposition, because the hatching nymph is usually of a certain size, shape, color, and the trout will often stick to that one type in their feeding. If it is an Olive Dun you will usually be required to use an Olive Dun nymph of the same size or a little smaller." Refreshingly, the caddis received its due as well: "The Hare's Ear is especially good with or without the gold rib, when the sedges are on the water and hatching out. Then, if the fish change off to a diet of the mature fly which flies over and rides the surface of the water, it will probably be advisable to use a dry fly or fish a wet fly on the surface." (Here we can assume that Leisenring also imitated the emerger stage.)

THE LEISENRING LIFT

Were it not for Pete Hidy, Leisenring might have been largely forgotten. Hidy's editing of the original manuscript presumably added material on presentation or at least called attention to that aspect, thus rounding out the work. The Leisenring Lift is a term still familiar to many fly-fishers. Here's how Big Jim described the technique:

> Imagine we are on the water. I cast my fly up and across about fifteen feet or more above where the big trout is located, depending upon the pool and stream. The fly sinks to the bottom, progressing along naturally, as I follow it with my rod, allowing no slack in the line, but being very careful not to pull against it and cause it to move un-naturally. The fly comes straight down to him, bumpety-bump over the gravel and stones along the bottom with the current. Now watch the fly. It is almost to him, and would only have to travel about four more feet to pass right by his nose without his looking at it unless it can be made appear alive and escaping. At this point the progress of the rod following the fly is checked, and the pressure of the water against the stationary line and leader is slowly lifting the fly. Now the fly becomes slightly efficient or animated and deadly, and the trout notices it. The hackles or legs start to work, opening and closing, and our trout is backing down-stream in order to watch the fly a little more, because he is not quite persuaded as yet. Now you can see the fly become even more deadly. As more water flows

against the line, the fly rises higher off the bottom and the hackle is working in every fiber. It will jump out of the water in a minute, now, and the trout is coming for it. Bang! He's got it.

Note how similar (in substance) this description is to Sawyers. Big Jim goes on to reiterate that the "lift" is entirely passive: "I try not to impart any fancy movements to my fly with the rod. The water will do all that's necessary to make a fly deadly." The Sawyer and Leisenring presentations largely differed in the span of the upward rise. In Sawyer's case the elevated rod tip moved the fly upward only a matter of inches whereas the lift covered a much longer arc, rod tip raised from horizontal to high overhead, as the current lifted the fly ever higher. Rick Haffle describes and illustrates the lift in his *Nymph-Fishing Rivers and Streams* (2006), noting that "the important thing with this method is to strive to make your fly lifelike in action and present it where the fish are feeding." And practically, he notes that depth and current are critical situational factors wherein added weight might be needed (heavier hook, leader or line) to reach bottom and that in slow water the nymph will likely need some help from the rod tip in ascending to the mid water or above—a modern version. Leisenring said that, conditions permitting, he caught almost all of his fish with the lift although turning to other methods if necessary. Leonard Wright also recognized the lift and said it was a technique "only the wariest or sleepiest trout can resist!" (*Flutter, Skitter and Skim*).

Legacies: Stewart, Skues, and Sawyer

Leisenring must have known the literature, for Stewart's Black Spider was one of his favorites. "I have found W. C. Stewart's Spiders to be a deadly combination on every stream I've ever fished. If a fly fisherman presents them carefully he can soon acquire the reputation of a fish hog!" Those pliant, breathing hackles were the common bond.

There's little doubt of Skues's influence. Leisenring dedicated his book "To G. E. M. Skues, Master Angler and Talented Writer." The two had a longtime correspondence, and Leisenring included examples of his flies in the letters. Of course the Big Jim's flies and nymphs were very like Skues's patterns, translucent, fur-dubbed bodies with emphasis upon soft, pliant hackle. Leisenring's best all-around pattern was the Tup's Nymph; same for Skues, and with almost identical recipes. (The Tup's served to imitate the larva of the Pale Watery Dun on both sides of the ocean.) Leisenring adapted Skues' baby plate idea for studying larvae (his was opal glass) and also employed a cheese-cloth net to catch insects kicked up from weed or gravel beds by a col-league. Nor was he offended by a messy autopsy. Leisenring's approach to entomology was empirical; the real thing and its imitation should look alike, no need for Latin. Accordingly, he judged the color of vari-ous larvae while still alive rather than storing them in a fixative.

It would appear that Sawyer and Leisenring refuted Skues's belief that emerging larvae are relatively inert; however, there is the mat-ter of depth to consider. Robson points out that the argument with Halford involved bulgers and feeding trout that disturbed the sur-face when few duns were afloat, activities of the top water. He sug-gests that expanding the discussion to the deep or mid-column layers would have been "off topic," a different can of larvae (*The Essential G. E. M. Skues*). Skues certainly recognized that nymphs swam about quite actively above weed beds and surely must have understood the induced take and "Leisenring lift," but never commented on either.

The possibility of a relationship between Sawyer and Leisenring is intriguing, given the similarities in their techniques. Although Big Jim was nearly thirty years younger, the two men's angling careers overlapped, and while the Brodhead and Avon were quite different fisheries, Leisenring's other favorite is said to have been the Little Lehigh, a limestone stream. Most compelling, both carried on a fairly extensive correspondence with Skues. Whether he served as an intermediary in some way is uncertain.

Big Jim's soft-hackled flies were a sharp contrast with the sleek, bullet-like PT Nymph, gently weighted with copper wire. Leisenring didn't insist upon casting to sighted fish and was more amenable to prospecting likely-looking holds. He also tied many more patterns as required to suggest prevalent larvae as the season progressed and so, despite the central issue of animation, the differences between the two were significant.

"FLYMPH"

What's in a name? Leisenring often referred to larvae happily pro-gressing through the stages of maturation on the streambed, and to their imitations, as *nymphs*. He called imitations of larvae ascending through the water column to the subsurface wet flies, and the subsequent adults simply flies. In their co-authored book as well as in *The Masters of the Nymph*, Hidy explains that a Flymph is none of these. Not a nymph because the insect has begun the process of emergence by swimming toward the surface. Not a wet fly or fly (adult) because the insect has only partially hatched. (I suppose we'd use the term *emerger* today.)

Hidy's flymph was tied as a hackle fly with the same attention paid to body color, form, and translucency as Leisenring preferred. In order to sink his flymph quickly, Hidy made the fly beefier, tying the body and thorax thicker with water-absorbent dubbing. His Flymphs were nonetheless fished just a few inches below the surface, either using natural drift or swinging across a feeding trout's position on a long leader and a thin tippet. Hook caliber was adjusted according to the insect at hand, whether mayfly larva or caddis pupa. He insisted upon natural furs and liked best the various shades from a hare's mask. The Flymph also had a more substantial muff of hackle. Ideally the body became enveloped in a thin film of air with a tiny bubble trapped in the hackle, thus resembling gas collected beneath the shuck during ecdesis, reminding us of Gary LaFontaine's Sparkle Caddis Pupa with its shiny veil of Antron yarn. When the Leisenring lift is illustrated in pure form the presentation is to a sighted fish whereas Hidy enjoyed prospecting likely currents by swimming flymphs upward with his rod tip during caddis hatches. *Flymph* has pretty much vanished from the fly-fisher's vocabulary. The name isn't easily swallowed, nor the concept, and when you read the two parts of the book in sequence, there are relatively few differences between the Flymph and the wet-fly patterns or presentation. A bonus, though: Hidy compiled an interesting history relevant to imitating that evanescent stage between larva/pupa and adult, mentioning contributions by Pritt, Hills, Schwiebert, and others.

DID LEISENRING RECEIVE HIS DUE?

There's no question that Leisenring's skills were legendary among regional anglers, and the introduction to *The Art of Tying the Wet Fly* was written by the well-known tyer Reuben Cross. But on a national scale? A photo in his book shows Big Jim during a vacation in the "American West" with a string of ten fish, maybe thirty pounds' worth, as he supplied the needs of hotels along the way, so he did get around. In later years he received invitations to sample a number of waters. And with the exception of Marryat most of these notable anglers developed a considerable identity through their articles in sporting journals, whereas Big Jim was not a writer of this sort. Here is a testimonial from Gary LaFontaine: "James Leisenring, through his writings, strongly influenced both my fishing tactics and my fly-tying." For one, he decided against putting wing pads on his Deep Pupa based on Leisenring's preference for a slim silhouette but did note that the lift, to be effective, had to take place pretty much in the trout's face, making other segments of its underwater journey relatively inefficient (*Caddisflies*, 1983). Schwiebert concludes, "Our exaggerated reverence for the Dry Fly Method and for the trinity of great American Fly Fishers who adapted its principles to our tumbling eastern rivers led us to overlook Leisenring and the excellence of his wet fly expertise." At the time of his death in 1952, according to Schwiebert, Big Jim was penniless. Many years later a "Day" was dedicated to Leisenring in Allentown, where a park was built in his honor (*Nymphs*, 2007).

The Nymph, Westward Ho!
with Rosborough, Brooks, and Nemes

A westerner can spend only so much time in Sullivan County, New York, and environs without getting twitchy. Norris, Gordon, La Branche, Hewitt, Jennings, and Bergman: that's quite a sequence. At this point in the project I kept thinking about the big drainages in the Sierras and Rockies, new fishermen and different trout, flies, and tactics. This chapter was somewhat therapeutic for me and accordingly is dedicated to Oregon's Polly Rosborough and his fuzzy nymphs, Charlie Brooks and his depth-charge patterns on the Henry's Fork, and Sylvester Nemes, sliding his soft hackles down the riffled currents of the Madison.

Polly Rosborough and the Fuzzy Nymphs!

It's a fair bet that many readers will not have heard of Polly, yet according to Schwiebert, "Rosborough and his original research are perhaps the first major contributions to the theory and techniques of American nymph fishing since Leisenring" (*Nymphs*, 1976). There are several reasons why Polly may have been largely forgotten. He was a son of the Pacific Northwest and although well known for his exploits on the Williamson, Klamath, and Deschutes rivers, his fame was largely regional. Then his only book, *Tying and Fishing the Fuzzy Nymphs*, published in 1965, primarily attracted the notice of serious tyers. It offered much more on details of construction of this unique construction than anecdotes or discussion of presentations. You're reminded of Leisenring's book. Polly's casual mention of ten-pound trout does catch your attention, though. His creations, the Fuzzy Nymphs, got more attention than the man. As a faithful follower of the fly-fishing literature, I recall pieces in *Field and Stream* and similar sources, either mentioning or devoted to patterns such as the Casual Dress, Nondescript, and Near Enough. Nor was he a mayfly man, as his Green Rock Worm, Golden Stone, and scud patterns attest. Rosborough also came up with innovative streamer designs. Directly or indirectly, he helped draw attention to the Pacific Northwest and the many exciting opportunities this diverse region offers for the fly fisherman.

How to "Fuzz"

Well illustrated with black-and-white photos, Polly's instructions are plain and to the point. Most of us use a thread core for dubbing nymph bodies. Rosborough's first efforts were with seal fur, that wonderful, kinky shiny stuff of yesteryear. Grey was a favorite; cream seal could be dyed. The idea was to form a thin mat to which a film of cement was applied, the kind we used for model airplanes. When the mat became tacky, the mat could be rolled into a permanent tapering "noodle" form.

One day Polly was attempting to un-snarl a skein of silk floss by soaking the mess in soapy water. It worked so well that he tried the same with fur right off the belly of a muskrat. The fibers spread out so nicely that after rinsing and drying, he was left with a mass of matted fibers. He teased out a small quantity and rolled it into a noodle. With additional spinning he had a pre-formed, tapered body something like Leisenring's bodies dubbed on silk. The magic, though, was that, in action, some of the surface fibers partially separated from the felted body to create a translucent fuzz. Better still, that fuzz pulsed and quivered in the current. When using fine-fibered furs, he roughed up the felted bodies with a hacksaw blade.

As their names imply, some of his patterns were impressionistic, and others were reasonable simulations of the naturals. Terry Lawton comments that the Black Drake Nymph was considered to be the most effective (West Coast) imitation of that insect for many years (*Nymph Fishing: A History of the Lore and Practice*, 2005). Like Skues, Rosborough paid close attention to conformation, a tapered body, ribbing, and wing-case bulge, often darker than the body. Beard hackle completed certain patterns and also slender tail fibers. In summary, Polly's patterns incorporated the basic features found in effective nymphs, form, translucency, softness, dynamics (the fuzz), and realistic color.

How He Fished Them

That depended upon the "where," a river or meandering meadow stream. Polly generally preferred a floating line (a sink-tip for big water) and favored cross-stream casts, especially when bulgers were active, dropping the un-weighted fly a bit upstream. (In order to avoid a splash, if conditions permitted, he dropped the fly onto the bank, then twitched it off.) Polly liked to fish his nymphs quite shallow, just inches below the surface. "All nymphs require a small amount of agitation in order to bring them alive. I give an extremely short, choppy action to the rod tip all through the drift and with the tip elevated abut thirty degrees above the horizontal plane and pointed at the fly as it drifts downstream. This causes the entire surface of the nymph to flow and become alive." (Where have we heard that before?)

Polly valued the coordinating action of the tail, legs, and wing cases as well. I'd never heard of some of his presentation techniques and was excited to try them. Picture this one: You're fishing downstream beginning with what many call a short tuck cast. "Letting out line just a little slower than the current to create a bit of drag while agitating the rod tip will often bring a slashing strike. Browns are real nasty about this. They often have to be teased into thinking a big hatch is in the making." It was the "induced take" all over again although downstream, more frantic, and non-stop. Polly never claimed that he developed the practice of constantly manipulating the fly and credited a friend with showing him how to add extra action by incorporating the aid of his line hand. Regardless, his legacy incorporates the tactic of presenting a hectically busy fly.

You Wonder

How many tyers use the felted noodle today? The ongoing explosion of dubbing materials, downy to kinky, dull to "rainbow," and in an infinite array of shades may have superceded Polly's method. Rosborough was a professional, selling his flies and tying materials, and after the manner of Gordon, the offerings of local birds and beasts were welcome. The "eyebrows" of road-killed Great Horned Owls were essential for his Black Drake, but he warned that his supply was not for sale.

To what extent were his patterns sold nationally? Ted Trueblood included at least one in his personal nymph series, but my 1973 copy of a popular West Coast catalogue doesn't list any of the fuzzy nymphs. It's a good guess that many of us tried his method anyhow. My first edition copy has lost the front cover and is shamefully defaced with my notes and underlines. Like the other great anglers, Polly was an observer and a thinker, developing his theories and methods on the basis of trial and error. Born in 1902, he was honored by the Federation of Flyfishers in 1975 and later produced a video tape. I've seen photos of Polly at an advanced age holding senior citizen trophy trout. He was regarded as something of a character and an exceptionally tough old bird. He lived, or better fished, to reach ninety-five!

QUICK WITH THE NYMPH

One of Polly's associates, Jim Quick, deserves mention in the context of early descriptions of nymph fishing in this country. His book *Fishing the Nymph* was published in 1960. It's hard to discover much about Quick other than that he was an outdoor writer, a professional tyer, and frequented some of the same fisheries as Rosborough. The book contains a lot of fundamental information, although you have to dig through Quick's cracker-barrel prose to get at the rewards. Nonetheless, he knew the literature, frequently quoting Hewitt, and corresponded with Skues. The intent and, to an extent, the structure of the book were models for Gary Borger's far more elegant, readable, and informative *Nymphing: A Basic Book*, published almost twenty years later. (Borger credits Quick with the first mention of using white thread as a tippet extension to achieve maximal suppleness.)

THE CAPITOL WOULD HAVE BEEN WEST YELLOWSTONE

When the holidays are over and the January blues set in, I reach for Charlie Brooks's *The Henry's Fork* (1986). It's virtual fishing perhaps, but if I ration the chapters and read over some a second time, it's soon March when the ice is pretty much off our southwestern streams and the midges and smaller stonefly species are getting busy. The Henry's Fork of the Snake isn't just a river, it's a stage for the study of geology, western history, Indians, and fur trappers, not to mention a watershed full of trouty rivers and their bountiful tributaries. Brooks writes of both. Here's a tiny sample that jumps from the Green Drake hatch on the Henry's Fork to the struggles of the Nez Perce tribe, all on the same page!

> At this time, most every day during the hatch (emergence of flying adults) there will be 200 to 300 fishermen between Box Canyon and Osborne Bridge. They fish in perfect harmony, each following his or her hunches as to where to be, which way to cast, and whether to use a Green Drake or a Pale Morning Dun which also emerges, just to confuse things. Then toward the tail of the Green Drake hatch, to make things really easy for the trout and tough on the angler, along comes the Brown Drake (*Ephemera simulans*). But a jolly time is had by all.

There's a lot of useful information in that short paragraph, but before you can assimilate it all you're reading that in August 1877 the beleaguered Nez Perce camped nearby during their flight to Canada, hotly pursued by General Howard's troops. This delightful blending of only partially related content makes *The Henry's Fork* unique.

Fly-fishermen might argue that if a rectangle were cut out of contiguous areas of Montana, Wyoming, and Idaho, you'd have a wonderful new state. In 1986, when his book was published, Charlie Brooks could have served as governor, for he retired in West Yellowstone (the capital) after a career in the military. From the early days when Charlie was a ranger at Yosemite he was an angler for all seasons, rivers, hatches, and techniques and hardly restricted himself to the Snake River drainages. The Madison was his home river and the Firehole a dry-fly favorite.

Brooks is known in part for his successful preoccupation with the removal of very large fish from deep, swift currents. His book *Nymph Fishing for Larger Trout* (1976) offers separate chapters on the methods of Skues, Hewitt, Sawyer, and Leisenring. Better still, he discusses their application in sections of the Henry's Fork and other rivers familiar to many readers, matching Skues's technique to the placid Railroad Ranch, Hewitt's to riffled rivers, and Leisenring's to a variety of waters. Charlie's understanding of all aspects of nymphing, practical and theoretical, truly sets him apart. (Incidentally, Dave Whitlock's illustrations make a nice addition to the book.)

Brooks begins forthrightly by admitting that his own method was not always popular with many who tried it, typically in swift, boulder-strewn "stonefly water." Imitations of the hefty Salmon Fly and Yellow Stone were tied on #4, 4X long-shank hooks weighted with lead wire and fished on 0X and 1X tippets. For starters a fear of deep, intimidating currents had to be dealt with. Then came the matter of skill; intense concentration and lightning-fast reflexes were fundamental to success. That's not all. Learning to propel thirty-odd feet of high density sinking line and a ponderous nymph required a special technique. Brooks warns, "False casting a heavy weighted nymph is a form of Russian roulette with eyes and ears as stakes." False casting was proscribed for this reason. Finally, patience and endurance were essential. The Brooks Method meant covering long currents, yard by yard, slowly and carefully. My experience with his technique has been in lesser currents with lighter gear, and it was still hard work. (The results, although decent, were lesser too.) Perhaps reflecting his military background, Charlie laid out his battle strategy in five phases: One, "Position," meaning some twenty feet above the head of a deep channel while wading a bit to one side. Two, "First cast," meaning some fifteen feet further upstream, but only six feet out toward mid-river. Three, "Sinking," the fly drifts down, reaching bottom close in front of the angler. Four, "Control," line and fly are moving downstream, rod pointed above the line's entry into the water and lifted to allow only a slight droop in the line between rod tip and the surface. Five, "Fishing" from the across-current position until the line comes straight below. Brooks pivots facing across and down, gently lowering the rod tip and swinging it more downstream, still aimed at the line's entry point and maintaining a slight droop. After a pause while the bowed line straightens, a fresh, somewhat longer cast is made. The longer the cast the higher the rod is held, eventually arms over head during Phase Four, and then lowered during Phase Five. The resulting slack allows the fly to dead drift. (No additional line is released.)

We all have a personal idea of what constitutes a nice fish. Mine begins modestly at about two pounds. The trout Brooks was after were true wall-hanger trophies, the kind that attract a gaping crowd in a fly shop or bar. Brooks contributed an excellent chapter to *The Masters on the Nymph* (1979), where he reviews his method and discuses his favorite patterns. The Assam Dragon, meant to suggest a dragonfly nymph, developed quite a following among lake and reservoir anglers. The Riffle Devil, tied to suggest large, carnivorous beetle pupae, took some very large fish, both drifted deep and crawled using the hand-twist retrieve.

Brooks's patterns were not all "depth charges." A serious effort was made to identify whatever the trout were taking, as from gastric contents, followed by attempts to create a realistic impression—size, conformation, and color—hopefully matching the pre-hatch. Quite apart from the stoneflies, Brooks gives his pattern recipes for the whole array of aquatic insects, large and small, as well as shrimp and sow bugs. Some are his own recipes, others old favorites such as the Montana Stonefly, Zug Bug, and of course, the Woolly Worm. Discussing swift streams cluttered with snags and surfacing boulders unsuitable for a sinking line, he suggests a floating line and unweighted flies with sufficient weight on a long, thin leader to keep the fly down. A full chapter is devoted to upstream nymphing, so while Brooks says that 80 percent of his nymph fishing was with his "method," he used whatever approach best fit the situation at hand. Studying insects in connection with trout fishing is like living in a house with a million rooms. Opening a door to any of these only reveals the contents of that room—and some more doors. But as long as one is not overcome by the enormity of the task, and proceeds one room at a time, a lifetime of joyous learning lies ahead. "It is insect activity that begets fish activity." That sounds a bit like Vincent Marinaro; they were friends and confreres. Brooks wasn't all muscle.

The Brooks Brothers

Charles and Joseph were brothers, if only within the angling fraternity. Joseph may have been the better known. He was a prolific writer of books and articles, a banquet speaker, and a TV celebrity. While both men targeted big fish, Joe angled the whole world.

I can't resist a personal fish story related to Joe Brooks's *A World of Fishing* (1964). Years ago when I read the chapter on Argentina an image got burned into my visual memory. It was pair of photos showing Brooks holding an eighteen-pound Brown and a famed Argentine guide holding another, even more impressive, "probably the largest fish (trout) ever taken on a fly anywhere"—twenty-four pounds. Thirty years later when my wife and I were planning a trip to Patagonia, I opened the book again and there they were, this time even more memorable, for we were to fish the Chimehuin and some of the other rivers Brooks mentioned. Fast forward: One evening on the Rio Traful I traded my dry fly for a flashback nymph in the deepening dusk. Fishing across and down, when the line paused, I set into an unseen something that moved slowly on down river, ponderously, imperturbably, in the manner of a locomotive. By the time it stopped, well into my backing, the guide was shouting for me to climb out. (We'd be late for dinner, served at 10:30 PM in Argentina.) I felt the thing throb occasionally, then the line went slack. Not much of a story? Back in the headlights when I snipped off my fly the hook had impaled the largest trout scale I've ever seen, maybe not tarpon caliber but at least the diameter of a dime if not a nickel. I said, "Wow!" The guide grumped, "Where's the fish?" You know *that* was a Joe Brooks fish! (I was going to mount the scale, but it dried out and shrunk.)

THE TRIPLE-THREAT SOFT-HACKLE

A friend of mine once met Sylvester Nemes working down the long, deep riffles of the Barns Hole on the Madison. Nemes stopped long enough from his fishing to chat. As I recall, my friend was less impressed by Nemes's flies, he called them Soft-Hackles, than the way he fished them and the results. After hearing the story I bought Nemes's first book, *The Soft-Hackled Fly* (1975). This short paragraph really caught my attention: "The soft hackled fly tempts trout a great deal more often. It tempts bigger trout. And it rouses the rapacity of the most lethargic trout causing it to charge from great distance or depth. That's why the take, when fishing these flies, is so powerful and so extremely physical." That would light any fly fisher's fire.

His soft-hackled flies were tied without tails, a very thin, silk floss body that left the back of the shank bare, and long, sparse hackles (only a twist or so). Certain patterns featured a plumped-up thorax via fur dubbing, and several, such as the Partridge and Orange, Partridge and Yellow, and March Brown were much the same as in

Pritt's book. Nemes tied a Tups, soft-hackle style, as well. I liked that our trout would accept those old Brit patterns. Why not? They offered realistic form, a bit of translucency, subdued, natural hues, and that magic, breathing hackle.

Nemes focused upon rivers with plenty of size and long riffles and fished them downstream using the "greased line" mending technique, well known to salmon fishermen. To paraphrase, the presentation goes something like this: Standing well above the head of a riffle, a long-ish cast is made to the far side. The fly drifts down under just enough pressure to turn it sideways in the current, but without so much that it begins to swing across. This requires the adroit and appropriate application of mends. Eventually the fly is allowed to swing to the near side when the angler takes a few steps downstream and repeats the sequence. Nemes found the first part of the drift to be most productive and the final swing across, least effective. His mantra was "Fish long, wade short." For greatest effectiveness Nemes needed space for "greased lining" and his flies best represented the species of larvae (caddis pupae too) common to riffles. In fact he states that smooth-surfaced streams with pretty uniform current such as the Henry's Fork Railroad Ranch section and Idaho's Silver Creek, the chalk-stream style, are not well suited for his method.

Nemes eventually located in Bozeman, Montana, surrounded by soft-hackle opportunities, and later revealed that certain dry fly fishers reported success fishing soft-hackles upstream. Gary Borger's classic *Nymphing: A Basic Book* (1979) offers support: Actually, the soft-hackled fly will float superbly when properly constructed. The soft hackle gives an excellent impression of crumpled wings and legs. And then in 1991 Nemes's *Soft-Hackled Fly Imitations* opened up new vistas by designing soft-hackles, fished like a dry fly, to imitate midges, caddis, and a number of mayflies (Tricos, *Callibaetis*, *Baetis*, and the Green Drake). Trapped in the film, a floating, crippled, or stillborn nymph might be suggested, or even a spinner. Nemes included a series of most convincing color plates with a natural (Pale Morning Dun and *Baetis*) beside an imitative soft-hackle. And he offered recipes for a whole slew of specific mayfly imitations.

I have a lot of respect for Mr. Nemes. Pritt is not especially well known among fly fishing's pioneers, yet Nemes found him, liked what he read, and successfully adapted the patterns, presented downstream, in a special water type. As far as I know, this was all by means of his own ingenuity and experimentation. Later he and others went on to show how soft-hackles could be effectively fished in many ways in various waters to suggest a variety of insects. The popularity of the soft-hackles is hard to assess. While I won't leave home without them, it's easy to understand why the plain-Jane simplicity of the patterns might fail to compete in fly-shop bins with today's shiny bead heads and flashbacks. Does that matter? Similarly, if Nemes is regarded primarily as a resurrectionist, more power to him! ⟞⟝

Marinaro's Modern
Dry Fly Code

We were at a social event the other evening when some of the ladies got into a competitive game of "who has been where." You know. "Oh my dear you must go!" It became rather tedious, but then I thought, Don't we fishermen do the same? We used to call it one-upsmanship. When playing with the big boys you'll need the Madison, Big Horn, San Juan, and the Henry's Fork under your vest to even ante up. Alaska is a gimme. When things get international, Mongolia packs quite a wallop, but so do trips to New Zealand (better both islands) with Tasmania as a kicker. If jabbed with Chile, counter with Argentina. And when I'm trapped in a corner, I can usually bob and weave out of danger's way via the Test or a bit of Scotland or Ireland where I caught salmon *after* the gillie said I wouldn't! But when someone plays the Pennsylvania limestone card, streams such as the Letort Spring Run, I fold. Why? Vincent Marinaro wrote two books featuring those very fisheries. In my 1970 reprint of *A Modern Dry-Fly Code* he talked about fishing the Letort during a typical mayfly hatch: "The capture of one good trout in an evening's fly-fishing was quite an achievement. Two good trout was a memorable event. Three of them would establish an enviable reputation." Later he mentions 8X tippets and #24 hooks, this at a time when 5X seemed pretty fine to me and #18 pretty small. Given that Marinaro, Charlie Fox, and his many other associates were profoundly expert fishermen, my shivers of awe and self-doubt may be understandable.

The Cumberland Valley and the "Club"

Vincent Marinaro was a native Pennsylvanian, born in 1911 in a small town northwest of Pittsburgh. He earned a law degree coupled with a BS at Duquesne University and became a corporate tax specialist, spending much of his life with his family in Mechanicsburg in the valley where Gordon first wet a line. "All of it began when as a young man from the raw mountain country of western Pennsylvania I looked for the first time on the fair face of the Letort and neighboring limestone streams. I knew instinctively that these waters were very different from anything I had hitherto known. They had a rich, fertile look about them that contained the promise of great fulfillment in the way of fishing pleasure."

I got a feeling for what Marinaro might have been like from a photo on the back flap of *A Modern Dry Fly Code*. Handsome, with Italian features and wearing a coat and tie (shirt with cuff links), he is scrutinizing a fly through a hand lens. There's a slight frown. It was the image of an intently preoccupied fly fisherman—or tax consultant. I later learned that he was also an accomplished linguist and violinist. Marinaro was not a contemplative angler of the placid, resting-beneath-a-shady-tree sort in the Walton sense, but much more the hunter in "a constant state of excitement in joyous pursuit of his quarry." His love and admiration for the limestone streams was such that the sun pretty much rose and set over the Cumberland Valley.

Lee Wulff said that of all solitary occupations, fly fishing is the most social. He was right. We apparently *need* to congregate and commune. There are no initiation ceremonies, secret handshakes, or passwords, but ours are still loose-knit fraternities (happily increasingly co-ed). The Marryat-Halford alliance with esteemed colleagues such as Francis Francis and Henry Hall was centered at Winchester. Poor Theodore Gordon's tiny circle of intimates would hardly qualify, but when you consider the array of pen pals with whom he regularly corresponded, his was the largest group of all. Meanwhile La Branche and Hewitt certainly enjoyed the camaraderie at The Anglers' Club. And so it was that Vincent Marinaro and Charlie Fox were the founders of The Fly Fishers Club of Harrisburg. Their underlying belief was that the limestone streams of the Cumberland Valley in south central Pennsylvania were the American counterpart of the Hampshire chalk streams. Charlie Fox owned property on the Letort with a stretch of meadow water analogous to some of the famous chalk stream beats. Marinaro describes fishing watercress farms on the upper stream. A photo shows him, recently returned from England, pouring tubes of water from the Itchen and Test into his beloved Letort, a marriage of the waters. The club eventually counted 300 members and was actively involved in a variety of conservation issues. (A full account is found in *Limestone Legends*, compiled and edited by Shires and Gilford, 1997.)

When nature's varied creatures exhibit this "clubbing" behavior, some general benefit always accrues. For you and me the answer must have to do with the sharing of observations, ideas, experiments, and outcomes. I mention this because, while Marinaro was most certainly his own man, his fishing provides good examples of "sharing." His friendship with Fox was something like the Marryat-Halford relationship, their skills virtually equal. (Fox also authored several important books, *Rising Trout* and *The Wonderful World of Trout* among them.)

A MODERN DRY FLY CODE

The first edition of *A Modern Dry-Fly Code* (1950) somehow escaped wide notice. Crown reprinted the book in 1970, adding black-and-white drawings and color plates. Today *A Modern Dry-Fly Code* is considered one of the great American fishing titles of the century. I've wondered if *code* was meant to suggest rules or codes of conduct—the sort of thing that makes some of us bristle. In reality, Marinaro, having discovered a set of truths pertinent to the fishing of dry flies in his limestone streams, suggested that they should be codified and acknowledged by serious fishermen. That sounds a bit like Halford. The code, reflecting prolonged and meticulous observation and experimentation, was proffered, not in the form of fiat or decree, but in the interests of sharing. Further, the principles have general applicability and are not necessarily restricted to streams such as the Letort. As I see it, the code is primarily based on three partially related theses.

1. The creation of a *realistic silhouette* of the insect being imitated is by far the most critical flytier's goal. As long as this is accomplished, the simpler the pattern, the better.

2. Minute insects, many of them terrestrials, constitute an important and generally unrecognized component of the trout's diet, and hence an overlooked opportunity for the fly fisherman.

3. The third has to do with spatial relationships between trout and fly (or insect), beginning when the fish first anticipates its prey and concluding when there is either a take or a refusal. Difficult to summarize, the implications for a fisherman are highly practical, as discussed below.

As to ethics, Marinaro, although a dry-fly man himself, referred to Halford as the "high priest of a puritanical cult" and wanted no part of the old argument. *In The Ring of The Rise* (1976), he wrote: "Generally it is unprofitable and decidedly harmful to cast the dry fly on fertile water containing well-fed fish that have not come up to a hatch." Under those circumstances it is far better to fish the wet fly, and for those who like to use it there are tremendous rewards for the skillful performer. (He nonetheless puzzled over why first-rate wet-fly fishermen impose themselves on these first-rate dry fly streams.)

In The Ring of the Rise

Marinaro's second book is a reflection of untold hours spent photographing trout and their antics from a blind on the banks of the Letort. Portions originally appeared in *Outdoor Life* and other periodicals. In good part the book substantiates, through these remarkable photos, what Marinaro told us earlier about a trout's reactions when taking an insect on the surface and also a trout's perception of a floating insect (or artificial). He further discuses the challenges and rewards implicit in the imitation of minute mayflies, ants, beetles, and other creatures and illustrates again the central roll of silhouette. As such, *In the Ring* serves as a delightful proof and extension of the elements of the *Dry-Fly Code*. The books should really be read in tandem. "There are those whose zeal and curiosity will lead them further and further afield in an effort to penetrate the murky gloom that obscures many aspects of fly-fishing, to dredge up bits of knowledge and to pry loose a few of the delicious secrets reluctantly yielded by a grudging nature." *In the Ring of the Rise* is all about penetrating that gloom and prying loose those secrets. (The cross-referencing between the two books is so extensive that I won't necessarily attempt to identify one or the other.)

The Jassid and Pontoon Hopper

There was a time when the Japanese Beetle provided the most exciting sport of the season for fly fishermen in the Cumberland Valley—and the most devastating destruction of many green things for farmers. The dichotomy was such that when the infestation faded, fair-minded fly fishermen had to admit to mixed feelings. Still there were many other, smaller Coleoptera that regularly flew or fell into the streams. The problem was that these mini-beetles were frustratingly difficult to imitate to the trout's satisfaction. Marinaro experimented unsuccessfully with dubbed-body beetles (given that the insects were thick little fellows) before he happened to mount a shiny black-and-white jungle cock feather (called a nail) on a hook, otherwise bare except for a palmered hackle, trimmed off top and bottom. The silhouette was spot-on, while the hackle legs stabilized and kept the feather flat in the film. The response from the trout was "prompt and electrifying." He called it "the first effective beetle imitation and the first breakthrough in a long series of breakthroughs that opened the door to a new and utterly fascinating kind of fly-fishing." Silhouette plus simplicity! He named his invention the Jassid. Jassids are leafhoppers and they come in all sorts of sizes, colors, and appetites for various grasses. The pastured banks of the Letort were teeming and the Jassid became immensely popular. It's still a limestone staple together with the Letort Cricket, credited to Ed Shenk, another famed limestone angler. Marinaro graciously said that the Cricket was "regarded by many as the most killing fly of modern times."

The Pontoon Hopper, based upon the principle of sillhouette, was the result of a cooperative effort with Marinaro's friend Bill Bennett. The quill of a turkey tail feather cut off near the tip served as a rounded, tapering body. (The quill's open end was plugged with a cork stopper.) Smaller quills became stabilizing pontoon legs lashed to the sides while a treble hook (with two of its members amputated to form a cradle) was fixed to the underside via the same lashing. Bennett added moose mane antennae and eye dots for a bit of panache while the thorax, wing, and upper legs were painted olive, green, and brown. Marinaro named the pattern after Bennett and said that "its virtues far exceeded anything else being used then or now." (Just a hint of Hewitt there.) Bass, too, fell to its charms by the score. Marinaro and Bennett came up with a similar Green Drake

Spinner imitation, this time using a hollowed-out porcupine quill tied on a short-shank hook with a thin layer of dubbing to simulate a thorax. It proved the undoing of selective browns. The Pontoon Hopper and porcupine spinner may have been more lure than fly, but in this day of the Chernobyl Ant and other foam-bodied monstrosities, we'd best not quibble.

Wings Ahoy!

Spinners and terrestrials lie flat in the film, in or out of the window, their bodies silhouetted. But the trout's first view of a mayfly dun is of its tall wings as it nears the "window sill." Today we speak of features in a fly that serve to "trigger" a feeding response. Marinaro saw profiles in just this way and believed that trout fasten their attention on only one particular trigger at any one time. (Hewitt had the same idea.) Since the wing image seemed a perfect candidate, he exaggerated the profile as to height and breadth. I'm reminded of a cartoonist dramatizing a particular facial feature such as a prominent nose.

A cut or trimmed and shaped wing from the webby part of a broad neck hackle was best suited. Flat and durable with stiff fibers, the flexible rib prevented the fly from spinning and twisting. Depending upon the insect, either a blue-gray absorptive shade or a more reflective yellowish white feather was chosen. Marinaro's dun imitations caught on so quickly that two club members even designed and perfected "wing cutters." Accordingly, if the wing is "it," hackle color must not be terribly important. Indeed he said that hackle should be subservient, helping to emphasize wing structure and blending in for color. (Here he and Hewitt had opposite views.) That trout mistake hackle for legs he thought preposterous. Marinaro believed that trout take a spider's hackle instead for wings and that this, rather than light pattern or movement, was responsible for the spider's success. He commented acerbically, "Mr. Hewitt invented a better fly than he knew." As was his custom, Marinaro backed up his preoccupation with wing structure with evidence: "During a heavy hatch of Henricksons, a particular trout was observed picking off virtually every dun arriving along a well defined current line. A colleague positioned himself at the top of the run where he caught a dun, plucked away its wings and released the body into the drift line. Over thirty of the impaired insects ran the gantlet without attracting a rise from the busily feeding fish."

In fairness, although it's unlikely that Halford understood the trout's early view of the wing tips over the "window sill" as later demonstrated by Mottram, he too greatly exaggerated wing structure. The cover of Tony Hayter's biography displays two flies with wings as prominent as seen in Marinaro's Green Drake, similarly displayed on the cover of A Modern Dry-Fly Code. Illustrations of Halford's "new patterns" in Modern Development of the Dry Fly indicate that the "big wing" was regarded as an essential.

Marinaro's insistence upon the central role of wing structure in imitations of duns (and spinners too) lived on. Doug Swisher and Carl Richards brought the concept to our attention (Selective Trout, 1971) and, going a step further, tied the hackle "parachute" or eliminated it altogether (the "Sidewinder"). My collection of fly fishing magazines from that era onward is dotted with articles celebrating big wing patterns, cutters, and wing burners. Over time wings have seemed to grow a bit smaller, but the hair wings of Al Caucci's Comparaduns are a dominant feature and have became increasingly popular.

Sailboat wings and mayflies go together. Marinaro wrote of the anticipated seasonal hatches almost poetically. Referring to the Hendrickson: "No sight is prettier than to see a string of these creatures come sailing along, wings erect and proud and slanted at a rakish angle, bodies rocking mildly with each little wavelet, causing the tall wings to tip and bow, now this way, now that as they turn with the current." The blue winged sulphur, bountiful and successfully imitated as both the dun and spinner, is the subject of an entire chapter in A Modern Code. From the first week in May until the second week in June, hatching each day, they are the great staff upon which every limestone fisherman leans for the bulk of his dry-fly fishing.

Thorax Only

Marinaro pointed out that where a trout sees the fly with reference to the window is important "but also how a fly should be designed to meet his views." In that regard his thorax modification of hackling a dry fly serves the same purpose as the outsize wings. The hackle is anchored at the midpoint of the hook's shaft rather than toward the eye. This almost eliminates that portion of the hook shank ordinarily occupied by a fly's body. Similarly, the sharply up-curved abdomen and tails of the natural are free of the surface and thus essentially invisible while the fly, natural or otherwise, is in the mirror. Two somewhat short-fibered hackle feathers are "X'd" fore and aft of the wing base. As a result the hackle provides "maximum support with stability." Tails (long hackle fibers), widely split, act as governors or outriggers to prevent capsizing. Far more important to the trout is the presence of the forebody or thorax. Without exception this part of the body hugs the surface of the water closely, oftentimes touching it. The thorax is sometimes part of the light pattern beyond the circumference of the window and is most important in the window itself.

All this Marinaro proves through photographs taken looking upward toward the surface in his "slant tank," a modification of Hewitt's apparatus. In one the image of a live dun is seen floating beside a thorax imitation in the "mirror." Both appear as "shapeless blobs," very similar except that the hackle creates a larger and streakier light pattern. In another set of photos the imitation is shown gradually approaching the window's edge. Sure enough the very tips of the splayed wings begin to peek above until, when the fly reaches a pristmatic band marking the very edge, the wings stand tall while the thorax blob remains indistinct. (Mottram also described this band.) Marinaro tended to regard light patterns created by whatever insect as an early invitation to take a more critical look and final decision when the insect edged into the window.

The crux: He found that trout consistently make their rise when the target is just at the upstream edge of the window where images are still quite blurred. (I'd always incorrectly assumed that the attack took place when the insect or fly was centered in the window, free of refraction issues.) Marinaro confirms: "The dun, riding lightly above the surface film, is never clearly defined at the point where the trout sees, inspects and takes the insect. If the fly breaks through the film, *you are showing the trout too much.*"

At the time, and to an extent still, the thorax style was quite popular for both duns and spinners. Given the importance Marinaro attached to silhouette and structure we can see why he depended upon only a small number of imitations. There are good reasons for the success of the one-fly man at least 50 percent of the time when fishing to duns, if his one fly in different sizes has a blue or gray dun wing. He might be even more successful if he had one more pattern with a pale or yellowish-white wing to accord with another large group. Marinaro felt that the color issue remained unresolved and that a single insect species might display an array of colors and shades. Discussing the practice of tying imitations of male and female Henricksons with different body colors, he said, "There is no good reason for such a practice and there are a lot of reasons against it." (The thorax hackle lifts what little body remains off the surface in any case.) It's worth noting that Gordon's Catskill-style dry flies are in some ways the antithesis of Marinaro's. Black-and-white photos of flies tied by Gordon appear in Paul Schullery's *American Fly Fishing*. The sparse hackle is attached immediately behind the eye of the hook and in front of a very thin body that has virtually no taper.

Spinners

Marinaro had as much affection for spinners as for the duns of a species, sometimes more. He often found the heavy Green Drake hatches disappointing in that the trout seemed unaccountably uninterested and aloof. Perhaps an earlier surfeit had left them listless. The subsequent spinner fall was more entertaining. Trout could be found rising eagerly to the remnants of the preceding evening's spinner fall in eddies and along banks, a prelude to the fresh hatches to come later in the day. (Spinners quickly drown and wash away in the swifter waters of freestone streams.) Wing profile and associated light pattern were also key to spinner imitation. Anything that breaks the surface film is no longer obscured by the oblique rays or the diffusion above the film. Accordingly, spinners, large terrestrials, emergers, and rising nymphs are extremely well defined as to color, form, and parts. When the live naturals were viewed in the slant tank: "I was amazed to see the transformation in the appearance of the wing immediately after the spinner fell to the water. In every instance the wings took on a brilliant, translucent aspect with long, clear, colorless streaks." (Skues made a similar observation referring to the "spun-glass" brilliance of the spinner wing, viewed from beneath the surface.) Studying the wing structure of the naturals, Marinaro realized that "this effect was created by the folds in the wing, acting as water traps to form light condensers that gathered light above the surface and transmitted this intensified light to the trout." Bundled hackle fiber wings had this effect, but only if the fibers were widely spread or splayed, thus trapping water and forming long, brilliant light condensers. Marinaro's "Unfinished Fly" solved the problem.

Five or six turns of hackle, taken at mid-shaft, were fanned out on either side, thus widely spreading the individual fibers. The hackle wings were cleverly bound into two lateral bundles rather than being trimmed off, top and bottom. Classically spinner wings lie flat in the film, but Marinaro had observed that some species maintain the wings in a partially erect posture for some time (the Drake for one) and this the tier should honor. Still others such as the Henrickson fall sideways with one wing standing up and the other in the film. Here his imitation was tied with splayed wings and little or no hackle. Unbalanced, the fly fell on its side.

Stupendous Green Drake hatches were expected on many of the northern limestone streams such as Penn's Creek and Spring Creek where Marinaro's description of the annual convergence of hoards of anglers was much like Halford's. However, the big insects weren't found in the southern streams. During the 1940s Fox and Marinaro participated in attempts to introduce a brood stock with some early hint of success, but the Drakes failed to take hold. (Hewitt's sour riposte suggested that they went about it the wrong way.)

The Hanging Emerger

Marinaro mentions that Halford's old favorite, the Gold Ribbed Hare's Ear, tied without hackle, could be effective fished in the film during a hatch. With this experience came the idea of a pattern style that would maximize the surface film's support. The Hanging Emerger was tied something like a spinner with long hackle, but on a special hook, heavy in the bend and light at the eye. Attached to the leader via a riffle hitch, the fly fell tail first and hung vertically in the film, a few hackle fibers and the eye of the hook above the surface. As he explained, "The water repellent properties of the ascending nymph combine with the gripping action of the elastic surface film to hold the dangling nymph securely until emergence is completed" (*In the Ring of the Rise*).

The "Minutae" (Marinaro's Spelling)

The scenario is too familiar. Trout rising steadily, no visible insects in, on, or over the water, and your best patterns ignored. So it was one day on the Letort for Marinaro and Charlie Fox. Frustrated, Marinaro lay down on a grassy bank, gazing out over the water. This brimming stream always looks overfilled in photos, about to burst its banks; thus the surface was likely close to eye level. Schwiebert wrote,

> Marinaro watched the slow current pattern slide hypnotically past. Some time elapsed in pleasant reverie before he was suddenly aware of minute insects on the water. He rubbed his eyes, but they were really there: minuscule mayflies struggling in their diaphanous skins, tiny beetles like minute bubbles, ants awash in the surface film and countless other minutiae pinioned in the smooth current. (*Trout*, 1978)

If fly fishermen are capable of epiphanies, he had experienced one. The next step was logical, Marinaro fashioned a fine-mesh net, holding it steady in the current. A thin line was "drawn" across the fabric at the level of the meniscus by the bodies or minutae. In addition he found some fair-size ants that had been drifting awash, only the tops of their backs in the film. He later wrote:

> Throughout all of my adventures with the insects of my home waters, I became increasingly aware of a special class of insects quite distinct from pattern or geneology. That class is distinctive in its very small size. Every order of insect is represented by extremely small species. There are many very small beetles, ants, jassids and mayflies to name only a few. I have created a special small-fly class because it creates special problems.

During the 1945 season he made a total of thirty-nine visits to the Letort, finding the trout taking minutae on twenty occasions. He recognized that various forms of pollution had diminished the mayfly hatches and accordingly, accentuated the importance of the terrestrials as a food form. Marinaro well understood that a placid stream flanked by meadows, such as the Letort, was a perfect setting for fishing to minutae and mentions two instances when a dead day was enlivened by a violent wind squall that drove the anglers from the water only to be recalled by a massive rise—a "wind hatch."

While he admits that trout feeding on "beasties" of the meadow seldom go more than a pound, he continues, "All forms of the dry fly have their particular attraction, but none of them begins to approach the ineffable charm of fishing with minutae." The special problems to which he referred were the lack of #20 and smaller hooks and gut delicate enough to present the corresponding "fly-speck" imitations. These were gradually overcome. A fellow club member made him the gift of some rare 8X gut of the highest quality, 4 oz. test, while well-tempered hooks clear down to #28s became available.

It's interesting that the "light pattern" was a non-issue with the very tiniest insects because they lacked sufficient weight to indent the surface film And when a trout sipped one of these miniatures, the dimple was hardly perceptible even to veteran limestoners. Marinaro describes a sort of wink picked up out of the corner of an eye. Related, he warns that it was not unusual to find trout after minutae "bulging," as if for nymphs, literally "humping" the surface, yet nymph patterns were totally ignored. He and Fox found that a tiny floater would pretty consistently solve this problem. A few patterns sufficed. He liked a small Adams as a general terrestrial lookalike. Others included a miniature olive dun and beetle and ant patterns in red and black. Among the minutae, Marinaro helped bring previously unrecognized *Baetis* species to the fly fisher's attention (*A Modern Dry Fly Code*).

Marinaro's Toughest Opponent

Local fishermen sometimes engaged in a pitched battle with overgrown, seemingly invincible trout. Marinaro once found a comfortable spot on a grassy bank ideal for rigging up his tackle. By chance an inviting eddy was close by, home to a large brown that fed steadily from the surface. "Hundreds of casts over many days were ignored. He had humbled me, this trout of the eddy. His refusals were as eloquent as the spoken word. So began my long vigil while the trout of the eddy continued to flirt his tail, and make his darting upward rises entirely unconcerned with my resentment and my watchful waiting."

Finally Marinaro discovered that his tormentor was taking houseflies—exclusively! (A dung heap was discovered just upstream.) But the frustration continued when the fish "frowned" at his imitations. When dealing with an opponent such as this Marinaro went through a series of attempted imitations (the game), looking for any sign of interest on the trout's part (a nod). He called it the "game of nods." In this way he gradually added this and discarded that, ever refining his imitation, until the game was won. In this case a few shiny, trembling hackle fibers tied to simulate a housefly's ever busy "hand washing" front legs did the trick. "This was the sign that the trout of the eddy was looking for. From that day forward all of my houseflies wore antennae, never fewer than two, never more than three" (*In The Ring of the Rise*).

THE HIDDEN HATCH

The Brits called the *Caenis* mayfly the "white curse." Marinaro called it "the most fascinating of all the insects that trout eat and fly-fishermen try to imitate." He used the term *hidden hatch* to refer to this mayfly's very rapid transition from larva to dun to spinner, almost "in air." (Remnants of the shuck could be found still clinging to the spinners; hence the hatch of duns was "hidden.") His descriptions of fishing the *Caenis* hatch leave you weak in the waders. The formula: Imitations are #24s, 7X tippets, trout to three pounds or so in weedy, snaggy surroundings. Descriptions of the hatches, "a solid wall of glinting, shining, sunstruck wings extending for twenty feet above the river" reminded me of the familiar "smoke" of a Trico hatch. And like "Trico scum," when the spinners blanket the water the open-mouthed trout take them "slobbering" or as Frank Sawyer said, like a "gobbling duck."

Marinaro relished the manifold challenge, his fly grossly outnumbered by the naturals, the take nearly imperceptible, the set liable to the parting of tippets and the ensuing battle even more so. Go to 6X? He found that the "heavier" tippet pressed the imitation (typically that of a female spinner) too firmly into the film, resulting in fewer takes. Remedies are suggested for each problem. Marinaro and Fox necessarily mastered the most gentle of sets, the rod kept low and pointed behind the trout, allowing the current to pull the fly into the corner of its mouth! Marinaro's preference for subduing a good trout was by "nagging him to death," rod pointed at the fish and played off the reel. (It's of note that several patterns from Halford's first book were recognized as effective limestone flies, among them a successful imitation of the female *Caenis* spinner.) In his 2007 revised and expanded *Nymphs*, Schwiebert discusses how *Caenis* and *Tricorythodes* species have been lumped by taxonomists. Perhaps Marinaro's mayfly would now be classified as a "Trico"?

Rise Rings

Marinaro's great powers of observation, infinite patience, and skills as a photographer are fully demonstrated in *Ring of the Rise*. The lessons the fly-fisher learns can hardly be underestimated. For starters, echoing La Branche, Marinaro found that when a cast is made *into* a rise ring in a relatively leisurely current, you're likely placing the fly well behind the trout. He called this the "gravest error." Why? Because trout, rising steadily, as in a rhythm, choose an "observation post" from which they detect approaching insects in the current line. But rather than attacking when the fly enters their window, they rise toward the surface, meanwhile drifting downstream under the targeted insect. At some point, after careful observation, they either take or refuse, but in either case promptly return to their original post. This brief description fits what Marinaro called the simple rise, most often seen when the trout are feeding confidently during a heavy hatch. The "drift back" under the insect before the take is relatively short. Presumably the trout have seen so many of the species that they're confident.

There are two bottom lines that Marinaro stresses repeatedly. For one, this is a game of movement. The current moves and with it the fly and, as the trout follows, so does its window, changing in diameter according to the trout's depth. Then as mentioned, during a hatch when insects regularly arrive along the drift line, the trout assume a somewhat regular feeding rhythm. Implication: it's best if your fly approaches the "upstream window ledge" at the propitious moment when the fish is "timed" to rise. (Otherwise, spraying the fish with a rapid burst of casts might put it down; patience and observation are both virtues.)

Marinaro's "compound rise" includes a longer inspection drift (possibly twenty feet or more) presumably reflecting doubt on the trout's part. Another lesson: When we're overly anxious for a hookup and the fly's drift is repeatedly cut short in favor of making a fresh cast, we may be missing a lot of hookups. Fish out the cast! The "complex rise" is a further extension wherein the poor, undecided trout, wringing its fins, decides to let the insect drift past! But then, unable to resist the last chance temptation, the fish pivots, races after the departing prey, and nails it. Marinaro said that once the downstream chase begins the trout *invariably* takes before swiveling sharply to face back into the current for the voyage back to its post. The proofs come in the form of sequential sets of remarkable color photographs taken from a blind where Charlie Fox and Marinaro's son Sebastian assisted him. Diagrams are nicely matched to each rise form. The gentle currents, clear water, and accommodating trout of the Letort Spring Run made an ideal setting.

In the Ring offers the reader a special bonus, a chapter entitled "Anatomy of the Limestone and Freestone Trout Waters." The compare-and-contrast explanations are really excellent; succinct and better than anything I've read. Beginning with matters geological we're introduced to stream chemistry and many aspects of fisheries biology. Hewitt is gently taken to task for his comment about the dearth of English-type chalk streams in our country. Marinaro, perhaps using looser definitions, identifies examples in the northern tier of states from California eastward, with others in the Midwest.

CONTEMPORARIES

The formative years of A Modern Dry Fly Code overlapped with the later careers of Skues, Mottram, La Branche, and Hewitt. Marinaro expressed his opinions about what they had written and we wonder to what extent he might have been influenced by his older colleagues. Mottram in particular placed the same degree of importance upon silhouette, pointed out that trout rising to "invisibles" may be taking tiny terrestrials, and reaffirmed that a mayfly's tall wings first announce its impending entry into the window. Marinaro mentions Mottram's interesting ideas several times, but clearly explains how he had reached the same conclusions based upon nascent observation and experimentation. Gingrich believes that Mottram merely anticipated Marinaro. I agree, but also believe that Marinaro's accomplishments serve to validate Mottram's place in history.

When you read Hewitt and Marinaro back to back, far more similarities than differences emerge. One is crankiness. It makes sense; both were intent upon experimentation born of observation and speculation, both certain of the validity of their findings, and as a consequence, both highly opinionated. Marinaro's devotion to the Letort was so great that he couldn't resist disparaging other anglers who, by choice or most often circumstance, were not of the limestone persuasion. Freestone streams and the trout they contain were targeted.

"Freestone trout are often so lean and empty that sometimes their stomachs contain the most astonishing assortment of indigestibles— even sticks and stones." Speaking of streams that are "rarely productive of fly life and the absence of rises is the rule and not the exception," Marinaro continues:

> Those who like to fish dry at all times on such waters, need not concern themselves with an exact imitation for there is nothing to imitate on such waters. Indeed the use of a purported imitation in these circumstances is a mere affectation and rivals the practice of jousting at windmills. Trout in these waters were brought up on La Branche's book. Limestone fish have never read the book. If one fishes the latter very much or exclusively, it is easy enough to acquire a belief that any fly will do at any time and that there need not be a hatch to raise your trout, but that is because such trout are hungry and are likely to strain every nerve and fiber to get at something that looks alive and tempting.

Poor La Branche and his Pink Lady! Nor does Hewitt escape, recalling Marinaro's left-handed compliment regarding the spider: "The dry fly today is largely a mongrel of sorts wearing a coat of many colors under which are hidden a number of seemingly divergent views on imitation or non-imitation, as the case may be. It may ride quietly for a moment, then it may be skated or fluttered, then again it may be cast repeatedly to simulate a hatch, and very often it can be found bobbing and swooping upon the surface of brawling mountain torrents." There were clearly conflicts with the Catskill crowd. Maybe it was chauvinism run amok, but Marinaro's devotion was, after all, at the very core of his contributions.

The *OTHER* River

A man of impeccable character, Vincent nonetheless had an affair! There was another stream in his life, Michigan's Ausable River. Yes, he cheated on the Letort, even including a long, well-illustrated "love letter" chapter in his second book. "Of all the rivers I have fished none are friendlier or kinder to the fisherman." Although its substrate was sand, he thought the spring-fed Ausable more like an English chalk stream than any other and relished the challenges posed by its signature "sweepers," long pines leaning into or over the water from its banks. The Michigan caddis, actually a large mayfly (the "Hex"), posed difficulties that only increased his admiration.

Those Who Knew Him

Wanting to learn a little more about our subject, I was delighted to find *Limestone Legends*, published by Stackpole in 1997 and edited by Shires and Gilford. The subtitle *The Papers and Recollections of the Fly Fishers Club of Harrisburg, 1947–1997*, refers to nearly fifty papers given over that span by as many members and a list of memorable guest speakers. A photo shows Hewitt holding forth at the very first dinner held by the Harrisburg Fly Fishers in 1948 shortly before *A Modern Method* appeared. Sparse Grey Hackle warned fellow members of the Anglers' Club of New York that the new breakthroughs were happening, not in the Catskills, but in southern Pennsylvania.

I turned to a chapter by Joe Carricato, an outdoor writer for a paper in Harrisburg and Marinaro's close friend and neighbor in Mechanicsburg. Marinaro is portrayed as a man of moods, often dour, his affect phlegmatic. He disliked crowds and felt uncomfortable with strangers, yet on occasion, as when speaking to a group, a remarkable sense of humor was revealed. That Marinaro preferred to cast only to rising fish comes across plainly in his books. I wondered, though, if he was as dogmatic as Halford in this regard. Carricato confirms that if no fish were rising, Marinaro would simply sit bleakly and wait. Although he never questioned the effectiveness of the wet fly in expert hands, his approval of the method is another matter. Schwiebert said that they often corresponded, that is until *Nymphs* was published in 1973. He never heard from Marinaro after that (*Nymphs*, 2007)!

A number of these recollections had to do with hunting, a pastime Marinaro enjoyed equally with his fishing. He was a crack wing-shooter and expert collector and connoisseur of mushrooms. Carricato said that on any subject there were two views, Marinaro's and the wrong one. Here's an example: There were already a bunch of children in Carricato's family when his wife presented him with twins. Marinaro came by the house and watched as Mrs. Carricato changed the first baby's diaper. Then he took over, diapering the second infant, and showing how it *should* be done. She just laughed. Charlie Fox provides some additional insights into his friend Vince's character suggesting that "Vince wasn't real free with information" or his flies. (Much like Gordon.) In order to stay supplied with Jassids, Fox and another friend would "stalk" Marinaro, recovering flies from trees and also from trout broken off by Vincent's overly vigorous sets with his stiff rod.

Charlie Fox was apparently the exact opposite, outgoing and interested in everyone. It's said that the affable Fox, rather than the taciturn Marinaro, was the ringleader of the "Pennsylvania boys," as Sparse Grey Hackle dubbed them. A photo in *Limestone Legends* shows the Schwieberts, father and son, taking instruction from Fox on the Letort. The youthful Ernie, stroking his chin, looks perplexed. Fox, a long time activist in fisheries management and conservation, is also credited for helping Trout Unlimited and the Federation of Fly Fishers to introduce catch and release initiatives.

At a point where I'd nearly finished *In the Ring of the Rise*, my image of Marinaro was not altogether complimentary—a steely-eyed, high-tech, opinionated trout hunter, planning each attack like a diamond cutter, hammer poised. But then came a page or so concerning brook trout. During an expedition for salmon when the fishing turned sour, Marinaro looked for another diversion, finding a small stream deep in the woods.

> The little stream ran over clean, bright gravel, sometimes in little quiet pools, with an occasional shaft of sunlight playing on its surface, sometimes running down a gentle riffle, quarreling and talking to itself, sometimes dropping in miniature waterfalls. When I returned home from that distant and expensive fishing trip, did I bring back unforgettable memories of battling big, fresh-run Atlantic Salmon? Not so. The unforgettable memory was that of a ten inch monster brook trout and the Lilliputian battle in that sun-dappled pool.

Vincent Marinaro's writing is consistently clear and pleasing with bursts of prose you want to read over and again. As others have said, Marinaro wrote with "elegance." There's another aspect. If we can't judge a book by its cover or its illustrations (photo essays excepted), both of Marinaro's books are all the more pleasing thanks to the charming line drawings by Pearce Bates. Many pages are decorated along the margins, the illustration generally matched to the topic at hand. The subject might be a fly, an insect, fish, bird, depiction of stream structure, an angler in action, or a riparian scene. Some are decorative, others instructive (the steps in tying a Pontoon Hopper) and still others quaint (a housefly busily rubbing its forelegs). To me the marriage between the quality of the text and accompanying illustrations is among the most appealing in the angling literature.

While I purchased *Limestone Legends* as a reference, I enjoyed reading all of it, from fish stories (the thirty-one-inch brown from Big Springs) to assessments of stream quality, new fly patterns, and a potpourri of other topics. In the process I was fortunately disabused of the notion that the Cumberland Valley streams were like peas in a pod. Turns out that the Letort Spring Run, Yellow Breeches, and Big Spring are quite distinct. More important, the collection chronicles how a group of enthusiasts with a common interest lived together as a microsociety, sharing an Eden of fly streams. It can never happen again.

I thought that *Legends* would include some information about Marinaro's last years. Carricato only tells us that Marinaro became more and more reclusive following the death of his wife. He died of leukemia in 1986.

LEE WULFF, BIGGER THAN LIFE

*H*aving read a slew of Lee Wulff's pieces in various out-door magazines over many years and watched his TV specials, I thought I knew a lot about this famous angler. Jack Sampson's wonderfully complete and detailed biography, *Lee Wulff* (1995), showed otherwise. A few chapters into the text I scribbled "bigger than life" in my notepad. It turns out that others have honored Wulff with the same words. Paul Schullery said that his contributions were so far-ranging and diverse that they would burst the seams of a single chapter (*American Fly Fishing*). I nonetheless decided to attempt the task by tightening focus on how this American legend carried so many of the essentials forward from the 1920s into the modern era.

FROM ALASKA TO NYC VIA STANFORD AND PARIS

Born in Valdez, Alaska, in 1905, Wulff became an adept slayer of salmon while still in grade school. The family later moved to California where he was a three-letter man at San Diego State. Lean, muscular, and ath-letic, Wulff stood over six feet. Next came a degree in engineering from Stanford where after, to his parent's dismay, he moved to Paris to study painting. Wulff was anything but one-dimensional. Although his art earned praise in exhibits, the homeland and a more varied career called. Short on cash and prospects but brimming with confidence, he wrote to his mother, "I'm about to start out trying to sell myself." What Wulff had to sell included a way with the brush (illustrator) and with words (advertising copy) plus an engaging presence and excellent communication skills. While working for ad agencies in New York City in the late 1920s Lee married a fellow artist whom he'd met in Paris. Both found work in the city, and streams in the Catskills and Adirondacks began to beckon on weekends. There he met new friends including future editor John McDonald, Dan Bailey, John Atherton, and none other than Norman Rockwell. The difficult depression years caused the Wulffs to temporarily relocate in Louisville, eventually returning to the city, where Lee secured a position at DuPont Cellophane. Finding the commercial art world confining, Wulff increasingly entertained the idea of some form of fishing venture to support the family. This began with talks to club groups, some of which featured movie clips even at this early time. Wulff also produced pieces for sporting magazines, writing on speculation at this point in his career.

The Hewitt Legacy

I've the idea that Wulff and Hewitt were friends. At least Wulff was respectful. He caught his first "dry fly salmon" on Hewitt's Bivisible and listed it together with the spiders and skaters among his favorite patterns. A number of his opinions also resembled Hewitt's. Both were lukewarm hatch matchers. Lee said, "The trouble is that exact imitations may have the static look of the real things, but usually they don't act like the real things." Like Hewitt, he believed the abiding principle shared by successful flies was motion, whether applied by the angler or through the intrinsic qualities of the fly. He mentions the quivering, palmered hackle of a Woolly Worm (Dan Bailey's invention) and the surface skip of a dropper wet fly. And when a trout or salmon ignored a dead drifted fly he remarked that "By the slightest twitch of the rod the fly can be moved just enough to suggest life. Sometimes when the fly is floating and the leader is underwater, I quickly jerk the fly under the surface just as it comes to the fish." Wulff was intrigued by the Neversink Skater and bivisible but preferred patterns with impressionistic appeal, mentioning the Adams.

> On the Battenkill one of my most successful flies was a nymph tied with gray angora wool on a #10 hook with a couple of turns of peacock herl at the head and a touch at the tail. When I didn't

put the herl head on, it didn't seem to make any difference. At the time I was selling the standard dry flies for twenty-five cents each, and I tried to sell my best customers some of my Gray Nymphs, giving these customers a special price of fifteen cents each because the flies were so simple to tie. I couldn't sell them. My customers didn't feel that enough work had been put into the fly. Therefore, my customers weren't interested and felt the trout wouldn't be either. (*Lee Wulff on Flies*, 1980)

An especially enthusiastic streamer fisherman, he felt that the form and shape of the fly should jibe with that of resident minnows in accordance with a certain artistic license. Beginning fly tiers were encouraged by the story of Wulff's "Wretched Mess," a sort of a nymph/streamer hybrid with hackle legs in front and marabou legs behind, something like a frog. He tied the fly with whatever materials might be handy, and it caught lots of fish. His inventiveness went beyond flies. Wulff was unhappy with the long, heavy shooting coats fly fishermen wore in earlier days and in 1931 he sewed the first vest of the modern type, short, sleeveless, and full of pockets. Orvis used it as a prototype.

The Big Bad Wulffs

The inspiration for the first Wulff dry flies, Gray and White, came circa 1931, through an attempt to attract trout rising to Gray Drake and Green Drake spinners, the "Coffin Fly." "I wanted a buggier-looking, heavier-bodied fly, and I needed more floatation to keep it up. Commercial imitations were too sparse and liable to drown. Buoyant bucktail was the perfect material for the wings and tail." (Wulff said that he was the first to incorporate hair in a fly pattern.) Instead of expensive blue-gray Andalusian hackle, he preferred the dyed feathers from "a local rooster butchered in Brooklyn." The White Wulff sported cream badger and the bodies were of angora, gray or cream.

Lee was a "lumper" when it came to patterns, that is, the Wulffs were a family of similarly constructed flies, eventually seven in all, buggy, easy to see, high floaters, and durable. (Lee once released fifty-one trout on a Wulff without changing the fly.) Dan Bailey felt that the series could be profitably marketed and convinced Lee to name them after himself. The Royal in particular was an immediate hit, thanks in part to Ray Bergman's mention of the pattern in *Trout*. I've always liked this explanation for the appeal of showy patterns like the Royal: "Trout, like people, have minds and special preferences of their own. Some are conformists, some are not. I feel that trout have preferences that they never lose, and when a particular type of food comes along, they will go out of their way to get it even though they are feeding at the time on other food which is plentiful." (Lee's analogy was a weakness for fresh strawberries with thick cream.) (*Lee Wulff on Flies.*)

Spiders and Variations on the Theme

A pair of scissors and a few marker pens were essential tools for the tier. Let's say a spider was not getting the job done. The angler might "scissor" the hackle flat on the bottom, presenting the fly flat in the film. Or to better suggest a spinner only the side hackle would survive. At the extreme, a 180-degree "buzz cut" might suggest an emerging nymph. Lee trimmed bivisibles too, perhaps creating an ant profile by trimming a waist. Markers were applied to adjust shade and tint. Regarding skaters, Wulff loved the soft pliability of bucktail and mixed a ring of delicate hair fibers with another of regular hackle behind. His great favorite, the radical Prefontaine, was named after a friend. This pattern sprouted a single bucktail wing pointing straight out in front like a snout. When skated across rough water, it apparently rocked from side to side in a most provocative way. The modern Bomber steelhead patterns are a legacy. Wulff had a persistent fascination with catching big fish on small flies and short rods as honored by membership in the very exclusive "Sixteen/Twenty Club—a twenty-or-more-pound salmon captured on a #16 hook." The provocative Prefontaine was one of the keys to membership.

Flexible Flies, Tubes, and Plastics: A Nod to Marinaro

Artists find excitement in the exploration of new and different media. True to form, Wulff found a way to combine form and function with simplicity. Any pattern can be tied as a flexible fly, simply by attaching the materials to an appropriate length of soft plastic tubing and locking the materials firmly with adhesive. He used Pliobond. The term *flexible* might be understood in two ways. The bare tube, hook in place and secured, could be decorated with whatever arranged in whichever way, wings, body, legs, and tails. And the tube formed a soft, flexible body. Wulff claimed that trout held a tube-tied nymph in their mouths much longer than a conventional nymph.

Looped-hackle wings developed by famed tyer Andre Puyans were a favorite. He used them to add an additional flexible (active) feature to imitations of moths, grasshoppers, nymphs, and even skaters. The entrepreneurial angler/artist patented his Form-A-Lure Plastic Bodied Flies. Going a step further, metal molds of various insect body forms and shapes yielded polystyrene "positives" secured to the shank of a hook. Rather than gluing parts and pieces to the pre-formed carcass, the plastic could be temporarily softened with a solvent. The hair, feathers, and whatnot were then applied before the polystyrene hardened. Wulff knew that plastic-cast flies had failed on the market and was disappointed when his flexible creations never became popular. Probably too unconventional, he thought. He knew his ideas had merit, for his Surface Stonefly was highly productive for large salmon and trout as well—"the fly I would hate most to give up." The plastic body came with a vertical post and was tied parachute with a bucktail wing (*Lee Wulff on Flies*). Only the late Gary LaFontaine might be Lee's equal in the imagination he incorporated into his patterns, always reflecting observation and logical experimentation.

NEWFOUNDLAND AND A NEW LIFE

Wulff always retained a love for trout fishing, but when he caught his first salmon on a fly in 1933 his life took a different direction. Still working as an illustrator, he could see a future in sport fishing in the vast river systems to the north with their heavy salmon runs. But before the requisite experience and skills for guiding could be developed, the huge bluefin tuna of the bays of Newfoundland and Nova Scotia diverted his attention. This was competitive big-game fishing, and Lee loved those competitions, becoming proficient (or better) so quickly that he was invited to join the United States team in a match against the British Empire's counterpart. The publicity created great material for his fishing pieces. One was published in the Anglers' Club Bulletin. He had made some important contacts and Samson wrote:

"Lee Wulff suddenly became a name in saltwater big-game fishing." By this time his byline had appeared in many of the major sporting journals, *Field & Stream, Outdoor Life, Sports Afield*, and others.

Combining this credential with ingenuity and a good deal of chutzpah, Lee launched his real career. In essence he suggested that the Newfoundland Tourist Board employ him to write accounts of the dramatic bluefin tuna and also their many excellent salmon fisheries for American outdoor publications. Prudently he first did his homework, determining that our anglers would, in fact, be interested in such stories. It worked! Wulff leapt into the Newfoundland salmon fisheries, learning all the while if sometimes swimming upstream against various impediments. In an address to a local Rotary Club: "There is no comparison between the amount of revenue derived by the government for each salmon that is netted as compared with that gained from a tourist for each salmon he catches." There were trophy trout to be had too, sea-run fish, and Lee was having great success with and writing about his Wulffs. He always incorporated the details of his tackle into these accounts, and readers began to hear about the short rod, an early favorite, just seven feet and two and a half ounces, at a time when "salmon rods" measured almost twice as long. He later fished a one-piece six-foot rod that Orvis later reproduced as the "Ultimate."

THE CONSERVATIONIST

Although the salmon runs were massive, Wulff deplored the extent of the slaughter, remembering the results of similar assaults upon the silver salmon around Valdez, and made this a point is his talks. The government soon after put a limit on salmon. In addition he was working on the manuscript for *Lee Wulff's Handbook of Freshwater Fishing*, eventually published in 1983. This brings us to something of a controversy. Lee Wulff's most famous words ever, written in the introduction, have led some to credit him with inventing the ethic of catch and release. "There is a growing tendency among anglers to release their fish, returning them to the water in order that they may furnish sport again for a brother angler. *Game fish are too valuable to be caught only once.*" It's not worth arguing about, but Hewitt suggested and substantiated that trout could and should be safely returned to the water in *Telling on the Trout* in 1926. Let's give them both credit.

World War II

Wulff's first book on salmon, *Leaping Salmon*, was published just before World War II. While I've not read it, comments suggest that it served effectively to update Hewitt's *Telling on the Salmon*, from 1922. With the advent of the war the family (now with two young sons) moved to Sushan, New York. Lee was assigned to Special Services to develop a recreational fishery for the military at bases in Newfoundland, where he guided celebrities such as Generals George Marshall and Hap Arnold. Meanwhile he continued to work for the Tourist Board, ever expanding his knowledge of the region and of fishing techniques such as the riffling hitch (a method of causing the fly to slide across the surface, exciting otherwise uninterested fish). At the end of the war these long absences continued, and while Lee remained close to his sons, a divorce followed.

To the Skies

Wulff was at his entrepreneurial best when he suggested that the Piper Aircraft Company loan him a light single-engine pontoon-equipped float plane for the production of a thirty-minute movie, promotional for the company and of course for Newfoundland fisheries—and for himself. They liked the plan so well that Lee was paid with a J-3 Piper Cub of his very own even though he had yet to learn to fly! (That problem took only three weeks or so to solve.) "In 1947," Jack Samson wrote, "it was unheard of for a man in a tiny, single engine aircraft to explore these great reaches of wilderness alone—searching for a place to fly fish for Atlantic salmon. Only large military aircraft dared to cross the rugged terrain and to brave the freak winds of these remote lands." Now Lee used the Cub to transport his own clients to and from his camps, meanwhile prospecting for new opportunities from on high where he could see and practically count the salmon and large trout. His illustrated lectures became ever more popular as fresh fishing pieces flew from his typewriter to the editors. In 1953 he gave a series of talks to the Anglers' Club that were so well received that he was asked to become a member. He had arrived and was further honored several years later when he accepted a position on the board of the *Atlantic Salmon Journal*. In 1956 he pushed further north into Labrador, now in a more powerful Super Cub. Despite these adventures, the exciting if contrasting possibilities offered by the saltwater flats interested him as well. *The Atlantic Salmon*, his most important book by far, and still a major resource, was published in 1958. In the 1960s he signed a remunerative contract with *CBS Sports Spectacular* for a mixed series on hunting (at which Lee was experienced and adept) and fishing for saltwater species. With all this Wulff found time to become a founding member of the Federation of Fly Fishermen as well as the Saltwater Flyrodders of America and Director of the Atlantic Salmon Association.

JOAN

Many honors surely, but far more importantly in 1966 Lee met and fell for a world fly-casting champion at an ABC Sportsman's Show. When Joan Wulff gave a presentation to our fly-fishing club after Lee's death, she said it was mutual love at first sight. This radical condensation of Jack Sampson's biography takes us only about halfway through the work. The remainder is concerned with the Wulffs' highly successful fishing school at Lew Beach, New York, and how Lee and Joan teamed up in various presentations and television programs. Joan became a celebrity through her popular book *Joan Wulff's Fly Fishing* and her pieces on casting for *Outdoor Life* and the *Fly Fisherman*. Sampson shares accounts of the first marlin, channel bass, and sailfish Lee took on a fly. Several chapters are devoted to Lee's never-ending devotion to the conservation of the Atlantic salmon, brokering his time, energy, and standing toward this end at many levels, including the international. It's doubtful that any sportsman has been more honored. The annual Lee Wulff Award given by the Atlantic Salmon Foundation is one of the most prestigious.

Lee died in a crash of his Super Cub in 1991, age eighty-five. A comment by Lee seems reflective of his long career: "We will have long discussions on the merits of the old days versus the new, but whether we like it or not the old days are gone. Our new world is bright and beautiful—as well as different. We're building wonderful fishing close to home. We'll make all our good available water as productive as possible. The best fishermen will still catch the most fish. Laugh or cry over the changes, fly fishing will always be fun" (quoted in Paul Schullery's *American Fly Fishing*).

I regret that I once thought of Lee Wulff primarily as a showman. He had gone to great lengths to share his amazing fishing feats (such as hooking, playing, and landing a salmon by hand) with TV viewers across the land. Jack Samson too said that Lee's goal was to become known as a great American fly fisherman. I imagined a sort of latter-day Buffalo Bill. But Wulff was the real thing. His cowboys and Indians weren't hired. The buffalo weren't tame, and his pistols shot live rounds. He has most deservedly become an icon. "Ruthian," he could knock salmon way out of the river with his stubby rod (if he chose to use one at all) and like the Babe, sometimes called his shots before the fact. The great American anglers who had gone before seem small and timid by comparison. Of course they weren't, but still the latter part of the twentieth century produced marvelous technologies that supported an eclectic renaissance. Lee Wulff, through his long and dynamic career, more than played his part. As noted in the introduction, the great anglers in history have had an ability to communicate through the written word. Lee excelled in person as well. As he said, the sport may be solitary, but we fly-fishers are social.

THE EMERGENCE

*P*redictably, fly fishing has progressed in each of its many nooks and corners over the past seven centuries. From a practical point of view, tackle technology comes first to mind. Simply, our modern tools allow us to fish better and more easily. We can choose rod, line, reel, leader, and even wading gear to fit our individual preferences and to rather precisely address whatever challenge may be at hand. Nor does decent tackle need to be expensive. On the other hand, Marryat accomplished amazing feats with the earliest split-cane rods, and Gordon and Hewitt considered the capture of large trout on the finest gut an important part of their sport, so it hasn't been all graphite and fluorocarbon.

But haven't there been biological parts to the story as well? For instance insights into the relationships between trout and insect that helped explain old and frustrating mysteries such as surface-feeding fish that rudely refuse our floating imitations. I'd argue that the phenomenon of emergence, that evanescent blink in the life span of aquatic insects, went pretty much un-appreciated until Skues blurred the distinction between dry fly and wet. It's not that the act of hatching went unrecognized. Halford observed that rising fish during a hatch of Grannom were either taking the pupae or the newly emerged adults swimming frantically to shore. A true dry-fly man, insisting upon a high-riding fly, wings erect and prettily cocked, he wanted no truck with damp, wiggling flies. The difficulty with the emergence argument is that there's no accepted definition of *emerger*. (In fact, it's not even a word according to Microsoft.) Sawyer and Leisenring began with what might be called the "contemplative stage," the larva's furtive drifts and movements up through the water column and perhaps back down. They gave credence to fishing the previously neglected "deep nymph." Halford's "bulgers" turned out to be fish chasing down seriously committed emergers approaching the surface, while Skues fished his drifting Tups in the film. Gary Borger described five "micro-stages" in the final "birthing" event, defining *emerger* as an insect in the act of actually shedding the shuck prefatory to full expansion of the wings. Somewhat to my surprise, Borger indicates that Step 3 (the insect has pulled its head out of the shuck, followed by the legs) is well imitated by a Parachute Adams! The correlative description is convincing and yet, for all these years we've thought of the P. Adams as a dry fly! It's an interesting and perceptive article. I imagine Gary would define many parachute patterns as emergers ("Film Flies," in *Fly Fisherman*, May 2008).

Swisher and Richards Make the Case

"The fish are getting smarter" is an observation that far antedates Marryat and Halford. Anglers have tried through various measures to circumvent the "experiential intelligence" of their adversaries. In 1971 Doug Swisher and Carl Richards wrote, "Even before the recent deluge of fishermen, however, there was a need for new patterns and techniques to fool those selective risers" (*Selective Trout*). Their popular book described how they proposed to address that need through their discoveries on the stream and in a lab equipped with aquaria and sophisticated photographic capabilities. It was not necessarily their intention to concentrate upon emergers, but take this passage: "Imitations of the emerging nymph are probably the most deadly and effective patterns of all, and many fish are fooled throughout the season by this breed of artificial." In *Tying the Swisher/Richards Flies* (1980), they discuss mayflies drifting on the surface: "Most anglers, at this time, switch over to a dun imitation, either a hackle or no-hackle variety. We have found, however, that a *Floating Nymph* is quite often much more effective." The reader is also advised not to ignore that stage when the larvae are ascending:

This is the signal for the trout to position themselves just under the top layer of current. During this phase of the hatch, they are actually focusing their eyes downward, sometimes even tilting their bodies forward, so they a can better intercept the rising nymphs. This is when the *emerging pattern* is supreme. If we were confined to select one pattern, and one pattern only, to catch the maximum number of fish during hatch situations, that pattern would undoubtedly be an emerger type.

It was and is a pretty powerful sell.

Swisher and Richards subscribed to the classic Skues fur-dubbed body and prominent wing case configuration but with truncated, upward-slanted wing stubs as if just bursting from the case. In these books they went on to expand the imitational possibilities matched to the situation of the naturals. We met the Extended Body Nymph, the seductive Wiggle Nymph, the Floating Nymph, and the Cripple and Stillborn forms. There was a caddis pupa too. Another nifty idea, instead of waiting for a hatch, try these patterns as attractors! Swisher and Richards finally said it plainly in *Emergers* (1991). Here they further refined their work with new patterns and materials such as the film-hanging "duck rump" (alias *cul de canard*) feather.

Caddis Take Their Rightful Place

I'm convinced that fly-fishermen were tardy in fully appreciating the little moths, so overshadowed by the seductive mayfly, cavorting in the deepening dusk after we had quit for the day. Leonard Wright called it right back in 1972, titling his first chapter "The Fly That Fishermen Forgot." The "sedge" sat at the back of the dry fly fisherman's bus until the twentieth century was well into its middle age. Admittedly, dense Grannom hatches attracted notice way back. Ronalds included a recipe for an imitation, while Halford called the insect a "capital fly to fatten the fish," meanwhile complaining that the "seething mass of struggling flies just hatched" with the "fish boiling in all directions" often resulted in a catch of many small fish (*Dry-Fly Fishing in Theory and Practice*). And it's true that the Henryville Caddis was popular on the Brodhead in the 1920s. However Hewitt makes only casual mention of the insects while Jennings and Flick, dry-fly men, pretty much disrespected the *Trichopotera*. Most compelling, Ray Bergman, wise in all the ways of trout capture, paid no heed to the insects, even in the last editions of *Trout*.

By the mid 1950s Schwiebert devoted two pages to caddis in *Matching the Hatch* while Swisher and Richards spent about the same number of words on the subject (despite offering a number of interesting imitations)(*Selective Trout*). Accordingly, *The Caddis and the Angler*, by Larry Solomon and Eric Leiser (1977) was important in my education. It describes the life cycle, habits, and characteristics of twelve caddis families. Separate sections on the larvae, pupae, and adults include suggestions regarding presentation and patterns with extensive, illustrated tying instructions—all this in just over 200 well-thought-out pages. Soloman and Leiser should certainly be given credit.

However just four years later Gary LaFontaine's *Caddisflies* (1981) pretty much eclipsed the earlier work, developing and extending virtually every facet. Has any construction material, unique in its nature and application, had an impact as great as Antron, basic to LaFontaine's Deep Sparkle Pupa, Emergent Sparkle Pupa and Diving Caddis patterns? Gary describes DuPont's three-sided nylon (Antron) as a material intended to be used in rug manufacture. DuPont's idea was that incorporation of the material tended to mask dirt accumulation. Gary's idea was different. When a caddisfly pupa emerges it fills a transparent sheath around its body with air bubbles. These globules of air shimmer and sparkle as they reflect sunlight, creating a highly visible, triggering characteristic. This sparkle is the key to imitating the emergent caddisflies. Untold numbers of fly fishers have come to agree, further solidifying the importance of the emerger and reminding us that "all that emerges is not mayfly." LaFontaine further came up with simple pattern styles such as the Dancing Caddis and Diving Caddis that served to suggest a wide selection of different species without the necessity of more compulsively "matching the hatch." Thus Gary gave further credence to both the emerger and to the caddis.

Fly Fishermen Are All "Emergers"

I originally determined to end this project with the chapter devoted to Lee Wulff. He played a part in so many aspects of the sport that his death somehow seemed to coincide with the end of an era. When I thought back, though, there were other writers who helped shape my fishing in major ways and almost certainly, the careers of many others. When I bought Ernie Schwiebert's *Matching the Hatch*, hot off the press in 1955, it changed my fishing at once and forever. Eager but embarrassingly provincial, I was introduced to a vastly wider world. Written in Schwiebert's uniquely captivating style, as much as anything *Matching the Hatch* served as an invitation to learn, look, think—and hopefully to become a real fly fisher. To that point my flies had come from Wright & McGill in Denver, their "Wiltless Wing" trademark series. The labels had that mayfly ring: Ginger Quill, Blue Quill, Blue Dun and the like. But the clumsy construction and heavy, quill wings (snelled hooks too) defied any hope of much success outside of wilderness waters. Thanks to Schwiebert, the Wiltless Wingers took wing virtually over night, and thus began my attempts to copy his recipes for the major western species. I tied and carried them proudly in my vest as if I knew what I was doing. As reference sources I've depended heavily on Schwiebert's beautifully illustrated *Nymphs* (1973), his encyclopedic *Trout* (1978), and recently the wholly revised two-volume *Nymphs* (2007).

Gary Borger's *Nymphing, a Basic Book*, published in 1979, had much the same salutary effect for me as did *Matching the Hatch*. It went far beyond "basic" as I learned about an aquarium's worth of creatures, their antics, lookalikes, and means of imitation. The "strip tease" and other presentations were clearly explained and have paid off over and again. I appreciated that the line drawings in the entomology appendix, matched to the accompanying text, were not so detailed as to discourage. You could delve as deeply as you wished into the insect lore. And it would be hard to find a better, more functional list of simple imitations than in *Nymphing*. How many dry-fly men has Gary encouraged to stray from the ranks of the purist? He has arguably been the most prolific fly-fishing writer of all during the past thirty years. The issue of the importance of color in our imitations also brings Borger to mind. Ronalds, Halford, Skues, Gordon, and others could be called "colorists." It remained for Gary to resurrect the natural notion that color must matter through his Borger Color System wherein subtle gradations (147 color chips) in shade and tint are coded for easy reference. Although more comprehensive and scientific, the intent of the System is the same as Halford's in *Modern Development of the Dry Fly*. Mottram, La Branche, Hewitt, and Marinaro found silhouette, light pattern, and "movement" more important than color, but we can become as much or little colorists as we please. No reason to argue!

This brings us back to LaFontaine and his off-the-wall, imaginative, and all-inclusive approach to pattern design. *The Dry Flies: New Angles* (1990) and *Trout Flies: Proven Patterns* (1993) have a prominent place on my shelves. As had so many others, he blended observation with ingenuity in designing experiments in search of solutions—the trout as final arbiters. All fly tiers are latent inventors, and Gary gave us strong encouragement. *Fly Fishing the Mountain Lakes* (1998) is a personal favorite that takes me back to college days and explorations of Colorado's skyline lakes. But these are not everyday experiences. Take caddis larvae that lose their neural buoyancy during periods of barometric change and float to the surface where trout "gulp them like grapes." He has a floating caddis larva pattern. Or how about flights of migrating termites that get "downdrafted" onto a lake from on high, creating a hatch of sorts? (I've see such a windfall of grasshoppers up at 11,000 feet.) As always, Gary combined strategy, pattern development, adventure, and humor in this slim, yet highly instructive volume.

CAUCCI AND NASTASI AND THE COMPARADUN

The 1970s saw contributions from Caucci and Nastasi as fundamental to today's sport as those of Swisher and Richards. Their books, *Comparahatch* and *Hatches: A Complete Guide to the Hatches of North American Trout Streams* introduced us to today's favorite and winningest style of mayfly imitation, the Comparadun, simple of construction, faithful to silhouette, and easily adaptable to color and shade. *Fly-tyer's Color Guide* afforded the reader an opportunity to match any hatch. The Caucci/Nastasi books boasted an array of excellent color photos, naturals and imitations, depicting the mayfly "Super Hatches," both east and west, with seasonal correlations. The structural modifications of dun and spinner, the "no-hackles and parachute hackled paraduns" were stressed. (Craig Mathews's deadly Sparkle Dun with its shiny, trailing shuck came along a bit later.) It was quite a leap from Jennings's pioneering *A Book of Trout Flies* and a number of other excellent "angler entomologies" have been published recently. New ideas and materials have surely made an impact, but it seems that it was, to a considerable extent, Swisher and Richards and Caucci and

Nastasi who crystallized the core principles of dry-fly imitation that their predecessors had left them. They showed that a few rather simple patterns that reproduce the silhouette, light pattern, size, and shade of the naturals will take us a long way.

It's hard to think of Dave Whitlock as a historical figure despite his innovative fisheries contribution, the Whitlock Vibert Box, way back in 1978. (The box is an incubator for fertilized trout eggs combined with a protective nursery for the fry, set into the streambed.) An exceptional artist and creator of many productive patterns such as Dave's Hopper and the Red Squirrel Nymph, Whitlock excels as a speaker and instructor in fly-casting and fly-tying. Several angler-friendly entomologies carry his name together with books and articles on a wide range of angling topics. Dave's articles on midges first attracted my interest in the minutae. Whitlock deserved a chapter, but heck, I still buy things from him. That's not history.

What Would They Have Thought?

A dry fly with a nymph dropper cohabiting on the same leader! Halford would surely have found it an obscene and unnatural practice. Skues, I think, would have been understanding, if not necessarily a proponent. The precise presentation of a brace of flies to a singular target in the chalk-stream setting might have been pretty uncomfortable. A century downstream the ploy seems to have become almost standard procedure. My guide friends say they have their clients fishing tandem roughly 90 percent of the time. Why? More often than not the f fish take the nymph, but the floating fly (often a sizable attractor) gives the anglers something to watch, hopefully maintaining interest and focusing their attention. Meanwhile, rigged properly, the nymph is presented effectively.

Maybe the idea isn't that radical. Picture Cotton and all the others artfully skipping a dropper along the surface, the submerged stretcher fly responding nymph-like. It seems that today's dry fly purist is the guy who won't go wet unless there's virtually no hope on top, and then only reluctantly. More power to him or her, that's what it's should be all about. I expect there are still some faithful Halfordians too, surely along the chalk streams. I hope so. The ethic is so pure, so pristine, like a high mass. There's appeal in that.

A Rounder Wheel

About all that's left is to try to make the fly fishing wheel a little rounder. We know that the future will bring high tech. One idea: The folks who produce fly rods have a tendency to animate their product. The rod, already ethereal, "comes alive" in your hand, responding to your every whim—almost a soul mate. Today there's plenty of room for miniaturized electronics in a rod's hollow butt. A tiny battery could power all sorts of communication systems, send, receive and store. Perhaps energy could be transported along the rod's matrix fibers and filaments, facilitating its performance according to changes in temperature, wind, barometric pressure, or even the angler's mood at the moment. Or when an otherwise controlled wrist begins to flop or tailing loops threaten, a warning/corrective sensation might be transmitted through the rod grip. Watch that! And you've got to give Mottram credit. Ninety-six years ago we didn't understand the double helix, let alone genetic engineering. Maybe we're on the edge of creating insects designed toward the angler's benefit. It works with sheep; why not a "Supersedge"? If trout could be toughened up enough to tolerate a live well, bass-type tournaments would surely follow. I hope not. From my perspective, the wheel is quite round enough as it is!

BIBLIOGRAPHY

Atherton, John. *The Fly and the Fish*. New York: Macmillan, 1951.

Behnke, Robert. *Trout and Salmon of North America*. New York: Simon and Schuster, 2002.

————. *About Trout*. Guilford, Conn.: Lyons Press, 2007.

Bergman, Ray. *Trout*. New York: Alfred A. Knopf, 1945.

Borger, Gary. *Nymphing A Basic Book*. Harrisburg, Penn.: Stackpole, 1979.

————. *Film Flies*. Harrisburg, Penn.: *Fly Fisherman*, Vol. 39. May 2008.

Bowlker, Charkes. *Art of Angling*. Ludlow, England: 1833.

Brooks, Charles. *The Henry's Fork*. Piscataway, NJ: Winchester Press, 1986.

————. *Nymph Fishing for Larger Trout*. New York: Crown, 1976.

Brooks, Joseph. *A World of Fishing*. New York: Van Nostrand, 1964.

Brown, J. J. *Fly Fisherman's Gold*. Lyon, Miss.: Derrydale, 1993.

Caucci, Al and Nastasi, Bob. *A Complete Guide to the Hatches of North American Trout Streams*. Guilford, Conn.: Lyons Press, 1988.

Cawthorne, John. *The Fly Fisherman's Entomological Pattern Book*. Accokeek, Md.: Stoeger, 2000.

Cotton, Charles. *The Compleat Angler*, Part 2. New York: E. P. Dutton, 1927.

Flick, Arthur. *Art Flick's New Streamside Guide*. New York: Crown, 1969.

Francis, Austin. *Catskill Rivers*. Guilford, Conn.: Lyons Press, 2005.

Gingrich, Arnold. *The Fishing in Print*. New York: Winchester Press, 1974.

Gordon, Theodore. *The Complete Fly Fisherman: The Notes and Letters of Theodore Gordon*. Edited by John McDonald. New York: Theodore Gordon Flyfishers, 1970.

Grey, Sir Edward. *Fly Fishing*. Winchester, England: Shurlock, 1975.

Hafele, Rick. *Nymph-Fishing Rivers & Streams*. Harrisburg, Penn.: Stackpole, 2006.

Halford, F. M. *Floating Flies and How to Dress Them*. Devon, England: The Flyfisher's Classic Library, 1993.

————. *Dry Fly Fishing: Theory and Practice*. Redding, England: Barry Shurlock, 1973.

————. *An Angler's Autobiography*. Devon, England: Flyfisher's Classic Library, 1998.

————. *Modern Development of the Dry Fly*. Boston: Houghton, Mifflin, 1923.

————. *The Dry-Fly Man's Handbook*. Lanham, Md.: Derrydale Press, 2000.

Hayter, Tony. *F. M. Halford and the Dry Fly Revolution*. London: Robert Hale, 2002.

Herd, Andrew. *The Fly*. Shropshire, England: Medlar Press, 2003.

————. www.flyfishinghistory.com/biographies.htm.

Hewitt, Edward R. *Telling on the Trout*. New York: Charles Scribners, 1926.

————. *Hewitt's Handbook of Fly Fishing*. New York: Marchbanks Press, 1933.

————. *A Trout and Salmon Fisherman for Seventy-five Years*. London: Robert Hale, 2004.

Hills, John Waller. *A History of Fly Fishing For Trout*. New York: Stokes, 1921.

Jennings, Preston. *A Book of Trout Flies*. New York: Crown, 1970.

La Branche, George M. L. *The Dry Fly and Fast Water*. New York: Charles Scribner's Sons, 1951.

LaFontaine, Gary. *Caddisflies*. Guilford, Conn.: Lyons Press, 1981.

————. *Fly Fishing the Mountain Lakes*. Helena, Mont.: Greycliff, 1998.

Fly Fishing the Mountain Lakes. Helena, Mont.: Greycliff, 1998.

Law, Glen. *A Concise History of Fly Fishing*. Guilford, Conn.: Lyons Press, 2003.

Lawton. Terry. *Nymph Fishing*. Mechanicsburg, Penn.: Stackpole, 2005.

Leisenring, James E. and Hidy, Vernon S. *The Art of Tying the Wet Fly and Fishing the Flymph*. New York: Crown, 1971.

Marinaro, Vincent. *A Modern Dry Fly Code*. New York: Crown, 1970.

————. *In the Ring of the Rise*. Guilford, Conn.: Lyons Press, 1976.

McDonald, John. *Quill Gordon*. New York: Alfred A. Knopf, 1972.

————. ed. *The Complete Fly Fisherman: The Notes and Letters of Theodore Gordon*. New York: Theodore Gordon Fly Fishers, 1970.

Migel, Michael and Wright, Leonard. *The Masters on the Nymph*. New York: Lyons and Buford, 1979.

Miller, Alfred. *Fishless Days and Angling Nights*. Guilford, Conn.: Lyons Press, 1971.

Mottram, J. C. *Fly Fishing Some New Arts and Mysteries*. Avon, England: Flyfisher's Classic Library, 1994.

Nemes, Sylvester. *The Soft-hackled Fly*. Old Greenwich, Conn.: Chatham Press, 1975.

_____. *Soft-Hackled Fly Imitations*. Bozeman, Mont., 1991.

Norris, Thaddeus P. *American Angler's Book*. Philadelphia: E. H. Butler, 1864.

Pritt, T. E. *North Country Flies*. West Yorkshire, England: Smith Settle, 1995.

Pulman, G. P. R. *Fly-Fishing for Trout*. London: Longman, Green, Brown and Longmans, 1851.

Ritz, Charles. *A Fly Fisher's Life*. New York: Henry Holt, 1959.

Robson, Kenneth, ed. *The Essential G. E. M. Skues*. London: A & C Black, 1998.

Ronalds, Alfred. *The Fly-Fisher's Entomology*. Secaucus, NJ: Wellfleet Press, 1990.

Rosborough, E. H. *Tying and Fishing the Fuzzy Nymphs*. Chiloquin, Ore., 1965.

Sampson, Jack. *Lee Wulff*. Portland, Ore.: Frank Amato, 1995.

Sawyer, Frank. *Nymphs and the Trout*. New York: Crown, 1973.

_____. *Keeper of the Stream*. Salisbury, England: Sawyer Nymphs, 2005.

Schullery, Paul. *American Fly Fishing*. New York: Nick Lyons Books, 1987.

Schwiebert, Ernest. *Matching the Hatch*. New York: Macmillan, 1955.

_____. *Nymphs*. New York: Winchester Press, 1973.

_____. *Trout*. New York: E. P. Dutton, 1978.

_____. *Nymphs*. Guilford, Conn.: Lyons Press, 2007.

Soloman, Larry and Leiser, Eric. *The Caddis and the Angler*. Harrisburg, Penn.: Stackpole, 1977.

Swisher, Doug and Richards, Carl. *Selective Trout*. New York: Crown, 1971.

_____. *Tying the Swisher/Richards Flies*. Harrisburg, Penn.: Stackpole, 1980.

_____. *Emergers*. New York: Lyons and Burford, 1991.

Shires, Norm and Gilford, Jim. ed. *Limestone Legends*. Harrisburg, Penn.: Stackpole, 1997.

Skues, G. E. M. *Minor Tactics of the Chalk Stream*. London: Adam and Charles Black, 1924.

_____. *The Way of a Trout With a Fly*. London: A & C Black, 1949.

_____. *Nymph Fishing for Chalk Stream Trout*. Devon, England: Fly Fisher's Classic Library, 1997.

Stewart, W. C. *The Practical Angler*. Devon, England: Fly fisher's Classic Library, 1996.

Sturgis, William Bayard. *Fly-Tying*. New York: Charles Scribner's Sons, 1940.

Wright, Leonard. *Flutter, Skitter and Skim*. Lanham, Md.: Derrydale, 2001.

Wulff, Lee. *Lee Wulff on Flies*. Harrisburg, Penn.: Stackpole, 1980.

Index